Advances in Industrial Control

Springer-Verlag London Ltd.

Other titles published in this Series:

Eric Allen and Marija Ilić

Price-Based Commitment Decisions in the Electricity Market

With 43 Figures

 Springer

Dr. Eric Allen
New York Power Pool, 3890 Carman Road, Schenectady, NY 12303, USA

Dr.Marija Ilić
Dept. of Electical Engineering and Computer Science, M.I.T.,
77 Massachusetts Avenue, 10-059, Cambridge, MA 02139, U.S.A.

British Library Cataloguing in Publication Data
Allen, Eric
    Price-based commitment decisions in the electricity market
    1. Electric power - Economic aspects   2. Electric power
    production - Economic aspects   3. Electric utilities -
    Decision making
    I. Title   II. Ilic, Marija
    333.7'9323

Library of Congress Cataloging-in-Publication Data
Allen, Eric, 1971-
    Price-based commitment decisions in the electricity market / Eric
    Allen and Marija Ilic.
        P.   cm — (Advances in industrial control)
    Includes bibliographical references and index.
    ISBN 978-1-4471-1162-7   ISBN 978-1-4471-0571-8 (eBook)
    DOI 10.1007/978-1-4471-0571-8

    1. Electric power — Purchasing — Decision making — Mathematical
    models.   2. Electric utilities — Rates — Mathematical models.
    I. Ilic, Marija D., 1951-   .   II. Titel.   III. Series.
    HD9685.A2A46 1999   98-41453
    333.793'2'0687—dc21   CIP

© Springer-Verlag London 1999
Originally published by Springer-Verlag London Limited in 1999

Typesetting: Camera ready by authors

69/3830-543210 Printed on acid-free paper

Advances in Industrial Control

Series Editors

Professor Michael J. Grimble, Professor of Industrial Systems and Director
Professor. Michael A. Johnson, Professor in Control Systems and Deputy Director

Industrial Control Centre
Department of Electronic and Electrical Engineering
University of Strathclyde
Graham Hills Building
50 George Street
Glasgow G1 1QE
United Kingdom

Series Advisory Board

Professor Dr-Ing J. Ackermann
DLR Institut für Robotik und Systemdynamik
Postfach 1116
D82230 Weßling
Germany

Professor I.D. Landau
Laboratoire d'Automatique de Grenoble
ENSIEG, BP 46
38402 Saint Martin d'Heres
France

Dr D.C. McFarlane
Department of Engineering
University of Cambridge
Cambridge CB2 1QJ
United Kingdom

Professor B. Wittenmark
Department of Automatic Control
Lund Institute of Technology
PO Box 118
S-221 00 Lund
Sweden

Professor D.W. Clarke
Department of Engineering Science
University of Oxford
Parks Road
Oxford OX1 3PJ
United Kingdom

Professor Dr -Ing M. Thoma
Westermannweg 7
40419 Hannover
Germany

Professor H. Kimura
Department of Mathematical Engineering and Information Physics
Faculty of Engineering
The University of Tokyo
7-3-1 Hongo
Bunkyo Ku
Tokyo 113
Japan

Professor A.J. Laub
College of Engineering - Dean's Office
University of California
One Shields Avenue
Davis
California 95616-5294
United States of America

Professor J.B. Moore
Department of Systems Engineering
The Australian National University
Research School of Physical Sciences
GPO Box 4
Canberra
ACT 2601
Australia

Dr M.K. Masten
Texas Instruments
2309 Northcrest
Plano
TX 75075
United States of America

Professor Ton Backx
AspenTech Europe B.V.
De Waal 32
NL-5684 PH Best
The Netherlands

Dedicated to my parents,
Owen and Candace,
my family, and God,
for their love and support.

E.H.A.

To my troubled
homeland.
With faith, love,
and hope.

M.D.I.

SERIES EDITORS' FOREWORD

The series *Advances in Industrial Control* aims to report and encourage technology transfer in control engineering. The rapid development of control technology impacts all areas of the control discipline. New theory, new controllers, actuators, sensors, new industrial processes, computer methods, new applications, new philosophies..., new challenges. Much of this development work resides in industrial reports, feasibility study papers and the reports of advanced collaborative projects. The series offers an opportunity for researchers to present an extended exposition of such new work in all aspects of industrial control for wider and rapid dissemination.

Deregulation of electricity markets is a worldwide activity with leading and distinctive developments occurring in the European and US markets. The immediate impact of these changes is felt in the higher levels of the power production hierarchy where tools to aid short-term economic decision making are needed. At the lower levels of the production control hierarchy the impact is to demand much more flexible unit operation so that the determined short-term goals can be met. Allen and Illic have produced this useful monograph on the decision algorithms needed to determine unit commitment profiles over the shorter time frame of hours and days. The method of dynamic programming has been used and the work covers modelling, cost function and construction, an introduction to dynamic programming and a detailed assessment of its use. The monograph comes complete with programs in Appendix D and with other appendices giving the necessary supporting theory. The research community in power systems engineering should find this comprehensive presentation an extremely valuable feature of the monograph, and it is a welcome addition to the *Advances in Industrial Control* Series.

M.J. Grimble and M.A. Johnson
Industrial Control Centre
Glasgow, Scotland, UK

PREFACE

The purpose of writing this book is two-fold: First, the book reports dynamic programming-based methods for distributed decision-making under uncertainties. In particular, the type of decisions of interest are on-off, rather than continuous. The overall objective is to optimize expected performance; this is done in an intelligent way, since the strategy is changing as more knowledge is gained in real-time about the uncertainties. The decision-making described here is typical of distributed controllers in a large-scale complex system, and reflects the self-adjusting and self-optimizing processes of the distributed decision makers to the uncertainties created somewhere else on the system. Assumptions typically made about the nature of uncertainties are relaxed; instead, major uncertainties become state variables themselves whose models are developed and used in the dynamic programming formulation of the problem.

Secondly, solving these theoretical problems is essential for successful decision-making for selling and/or buying products competitively in physical and financial markets. Physical markets are particularly challenging where the inter-temporal effects play a dominant role; for example, an electric power plant could optimize its profit by maximizing its revenue, which requires selling when the price of electricity is high, and by minimizing the total operating cost of delivering the committed power by taking into consideration startup and shutdown costs and must-run time constraints.

As the electric power industry evolves from being a fully regulated, cost-based industry into functionally and corporately separate generation, transmission and distribution entities, it becomes essential to bid intelligently into short-term (daily) physical electricity markets as well as longer-term financial markets (such as the 18 month futures market). The supply/demand aspect of the deregulated industry is becoming fully competitive and, as such, requires commitment decision methods similar to the methods used in other competitive industries. This is a significant change compared to the coordinated least-cost scheduling of generation done in today's industry. Instead, the decisions are motivated by the individual profit/benefit maximization criteria for the anticipated market conditions.

Depending on the specific electricity market structure in place, the obligation to supply sufficient power so that the expected demand is fully met is not an explicit constraint. The excess demand leads to the increase in the price of electricity, and this has a self-regulating effect likely to ensure that sufficient supply is committed to meet the required demand. Dynamic efficiency, measured in terms of total social welfare over a finite time, is also optimized through the distributed decision-making described here.

This book concerns only tools for the newly evolving electric power industry directly relevant for decision-making by suppliers and consumers of electricity. It does not concern real-time operations and planning of the transmission and distribution systems necessary to facilitate power transactions between sellers and buyers. As such, the methods described in this book should be of direct interest to future generation companies — GENCo's–, load serving entities — LSEs– and, more generally, power marketers.

The book describes basic methods for the intelligent selling and buying of electricity so that power producers and consumers optimize their individual profits. At the same time, assuming no market power problems, it can be shown that these decentralized decision-making methods lead to near-optimal systemwide efficiency of the electricity market, in which total expected social welfare is maximized. This is the essence of competitive economics, and the methods developed in this book provide basic formulations for decision-making by future GENCo's and LSEs in the competitive electricity markets.

ACKNOWLEDGEMENTS

This book is a direct outgrowth of Eric Allen's doctoral thesis at MIT under Marija Ilić's guidance. The work was motivated through the active interactions with the industry which indicated the need for developing strategic bidding tools for supplying and use of power into electricity markets. Particularly useful was the input of the participants in the MIT Energy Laboratory's Consortium on Transmission Provision and Pricing. Additionally, the financial support by this Consortium, as well as by the Department of Energy, is greatly acknowledged.

The authors are also very grateful for having been in the academic environment which made the need for this work clear, as well as for the rich intellectual synergy necessary to attempt an interdisciplinary topic of this type. An early discussion with Pravin Varaiya provided excellent encouragement to proceed with this difficult subject. John Tsitsiklis, Francisco Galiana, Robert Pindyck and Yu-Chi Ho provided critical feedback to this work as part of their service on Eric's thesis committee. Eric particularly acknowledges the Dynamic Programming course at MIT taught by Dimitri Bertsekas as a critical background for this work.

On a more personal note, Eric also gives thanks to his parents Owen and Candace, his sister Debbie, his brother Scott, and all members of his family for their unending love and concern. Marija is very grateful to her family for the love and understanding of the effort required to produce this book. Both Eric and Marija thank God for all the gifts He has given them.

CONTENTS

CHAPTER 1
INTRODUCTION

The coming structural changes in the electric utility industry require that market participants change the procedure by which both financial and operational decisions are made. This book presents methodologies by which such decisions may be made optimally. To the best of the authors' knowledge, this book is the first text to present dynamic programming (DP) [1] methods for decision problems in power systems using a stochastic framework. Previous work [2, 3, 4] has addressed deterministic problems in the regulated utility environment. Recent publications [5] have focused on the deregulated electricity marketplace, including the topics of price modeling and prediction, futures market strategies, and risk management; however, optimization problems with inter-temporal effects that require DP-based solution methods have not been addressed.

This book focuses on problems requiring zero-one decisions (i.e. on-off) for which price is an important factor. The objective of the problem is to maximize the expected profit or benefit over a time horizon of many hours, days, or even weeks. Such problems are faced by electric generator owners as well as large industrial users. Although this book presents examples with only two control options, the techniques are generalizable to problems where several options may be considered. Since the profit during any hour in the future is a random quantity when viewed from the present, the problem is stochastic in nature.

Because price is a major component of the profit, price models are developed in this book for predicting future price changes as accurately as possible. A simple price model is obtained by treating the prices for all future times as a series of independent identically distributed random variables. Such a model does not need to include price as a state variable; however, strong correlations in actual price data between subsequent hours make this model very questionable. Usually, prices are modeled using a continuous-time process [6, 7]. However, since electricity markets will generally evolve on an hourly basis, a discrete-time model is much more natural for representing electricity prices; such a model fits in well with the discrete-time models of other problem dynamics, such as the status of generators. Such models all include price as a state variable. The decision problems being solved are similar to the investment problems of [6]; however, in [6] a Brownian motion model is used for price fluctuations. An examination of electricity prices shows that such a model is clearly inappropriate, and more detailed models, such as an auto-regressive moving average (ARMA) process [8], are much more suitable. The lack of Brownian motion behavior by electricity also renders many economic tools, such as the Black-Scholes formula for option pricing, as unreliable. This book expands on price models such as ARMA by making use of external factors, notably temperature, that are correlated with price.

Throughout this book, the unit commitment problem is developed in detail as a representative example of a price-based decision problem. Unit commitment is the process by which generating units are turned on and off. Unit commitment decisions are subject to many constraints, such as minimum up and down times. Unit commitment is usually solved over a time horizon ranging from 24 hours to a week. This book focuses on the unit commitment problem in transition from a regulated environment to one that is deregulated and market-driven. However, the concepts in this book are not limited to unit commitment and may be applied to a much broader class of decision problems.

With the development of a futures market for electricity [5], new opportunities for risk hedging become available, but techniques for optimizing profits are needed to achieve the full benefits. This book formulates decision problems for forward contract sales to minimize risk. A basic one-step problem is presented and solved. A more general problem formulation accounting for inter-temporal effects over a much longer time horizon is also presented, but a solution is not given because of the vast amount and complexity of information required, such as the expected value of the variance of tomorrow's forward market price.

This book expands on previous literature in several ways. Most previous work on unit commitment [2, 3, 4, 9], although accounting for the costs arising from inter-temporal effects, uses a deterministic problem formulation for a regulated industry. [6, 7] cover the topic of price-based investment decisions using both dynamic programming and contingent claims analysis, but conclusions are developed primarily for a Brownian motion price process. A very recent work [5] focuses directly on the deregulated U.S. power industry, including many price models and a study of electricity futures markets, as well as a general view of risk management techniques. However, decision problems with inter-temporal effects, such as unit commitment, which require DP methods are largely unaddressed. This book aims to fill that gap by providing solution methods for a stochastic problem formulation for decision problems in the deregulated utility industry using price models which make use of external information for improved accuracy.

The book begins by presenting a generalized formulation of the present unit commitment problem in Chapter 2. The complexity of the generalized problem clearly illustrates the simplifications that are made in order to make the problem tractable. In particular, the stochastic aspects of unit commitment are often reduced to a deterministic form, which drastically reduces computation but also leads to a suboptimal solution. Interruptible power contracts are also presented; these contracts are most often associated with the deregulation movement, but they can also be included in a regulated environment.

Chapter 3 describes the likely forms of the spot market under deregulation. Two basic market structures are presented, and the unit commitment problem for an individual power producer under each scenario is presented. This problem is one of choosing whether to invest a fixed startup or shutdown cost now in order to maximize expected profit over the near term in the presence of uncertainties [6]. The unit commitment problem under deregulation has many fewer decision variables, and consequently many more characteristics can be included. In particular, a stochastic approach becomes feasible. Additionally, under the assumption of a competitive market, the unit commitment problem for an owner of multiple generators may be solved by optimizing each generator individually; such an approach further reduces computational complexity.

Chapter 4 gives an overview of dynamic programming [1] which will be used as the principal tool for solving unit commitment. Dynamic programming is used in order to account for inter-temporal effects that arise when a decision in one time period affects the options available in subsequent time periods. Such inter-temporal effects are present in the unit commitment problem, which includes minimum up and down times for a generator and startup and shutdown costs [3]. If the generator is turned on now, then it must remain on for a fixed length of time and a startup cost must be paid. A decision to turn on now may increase profit over the current hour but reduce profits over the next several hours. Dynamic programming is a technique often used to account for these effects. By contrast, static optimization techniques such as [10] and [11] only optimize over a single time step and do not consider these effects. Only recently in [12] have inter-temporal effects been considered in the literature; this book continues study in this direction.

Chapter 5 gives a detailed formulation of the cost functions for an individual power producer's unit commitment problem. The cost functions are developed both without and with upper and lower limits on the power generated at any given time. Without generation limits, the expected cost is a simple function of the price mean and variance. When generation limits are included, the expected cost can still be expressed as an analytic function by introducing truncated random variables. A truncated random variable has a probability density between two limits, but it also has positive probability of being equal to either limit.

Prices at consecutive hours are not independent of each other. In order to obtain a realistic solution to unit commitment, a price process model for electricity is needed. In chapter 6, a representative price model for electricity is developed from actual price data. This model takes into account both the correlation between prices at consecutive hours and the hour of day and time of year. Outside factors, notably temperature, have a significant effect on the price of electricity.

After the problem model is developed, the next question is how to solve it. Chapter 7 illustrates several viable methods for solving unit commitment on a hypothetical numerical example. Enumerative dynamic programming is the primary solution method used in this book; however, an application of ordinal optimization is also shown. Ordinal optimization [13, 14] aims to find a "good enough solution with high probability," and it is a possible approach to handling computational complexity.

Finally, several extensions of the deregulated unit commitment problem are considered. Chapter 8 explores the use of forward contracts. Forward contracts can be used to greatly reduce the risk inherent in profits which depend on spot prices in the distant future. Chapter 9 expands on the idea of a reserve market. Two possible payment methods for reserve are shown. The strategies for an individual power producer when offered the opportunity to sell reserve are illustrated. Chapter 10 discusses possible effects of transmission congestion. On the one hand, the price model may be sufficient to reflect the effects of congestion on a power seller; however, a fixed limit on power sales may also be imposed.

While a practical problem in its own right, unit commitment is an illustrative one as well. The methodologies developed in this book may also be applied to optimal decision-making by various market participants in several different market structures. Because of the conceptual symmetry between supply and demand, it is expected that buyers and sellers will make choices in a similar manner. For example, an industrial customer may have large electric machinery, such as a loom, that is expensive to turn on and off and can not run for short periods of time. Deciding whether to turn such a machine on or off in response to price is the demand-side analog of unit commitment, and hence unit commitment solution techniques are directly applicable. Therefore, we expect new businesses such as the generation serving entities (GSE) and the load serving entities (LSE) to rely heavily on price-based commitment decision approaches described in this book. The recently published book by the power marketers [5] clearly describes the need for the type of such computational tools.

CHAPTER 2

THE UNIT COMMITMENT PROBLEM

Unit commitment is the process of deciding in advance whether to turn on or off each generator on the power grid at a given hour. In order to illustrate the changing nature of unit commitment as the electric power industry moves into a deregulated market, a broad form of the unit commitment problem for the present environment is first presented. An overview of present solution methods is given in order to illustrate simplifications which are frequently made, such as neglecting the stochastic aspects of the problem. Although interruptible power contracts are usually associated with electricity deregulation and the end of the "obligation to serve," they may also be implemented in the regulated industry, and therefore they are included in the following unit commitment description.

2.1 Unit Commitment in a Regulated Industry

The problem of unit commitment is to determine, on a given day, which generation units should be running to meet the anticipated system load and reserve requirements. Since generators can not instantly turn on and produce power, unit commitment must be planned in advance so that enough generation is always available to handle system demand with an adequate reserve margin in the event that generators or transmission lines go out or demand exceeds the expected amount. The units are chosen so as to minimize the expected total cost over the long term horizon. The costs considered include the cost of generation, the startup and shutdown costs for each generator, the costs of failing to serve loads in the form of insurance payments, and the revenue received for each unit of energy used by the loads.

Unit commitment decisions are made at periodic intervals. The time period between decisions will be referred to as a stage. The total cost incurred during stage k, denoted as C_k, is:

$$C_k = \sum_{i=1}^{N_G} \left[\int_0^{h_k} c_{Gi}(P_{Gi}(t))\, dt + u_k(i)I(x_k(i) < 0)S_i \right.$$

$$\left. + (1 - u_k(i))I(x_k(i) > 0)T_i \right] + \sum_{i=1}^{N_L} \left[(1 - R_{Li})I_{Li} - p_i \int_0^{h_k} P_{Li}(t)\, dt \right]$$

$$(2.1)$$

The indicator variable I of a conditional statement has a value of 1 if the statement is true and 0 if it is false. In equation (2.1), the first term is the cost of generation. The next two terms are the startup and shutdown costs respectively, followed by insurance payments for interrupted loads, and lastly revenue for power delivered. The definitions of the symbols in this equation are shown in Table 2.1. The state value $x_k(i)$ is positive if generator i has been up for $x_k(i)$ stages and negative if generator i has been down for $-x_k(i)$ stages [3]. The state transition equation is given by [3]:

$$x_{k+1}(i) = \begin{cases} \max(1, x_k(i) + 1) & : u_k(i) = 1 \\ \min(-1, x_k(i) - 1) & : u_k(i) = 0 \end{cases} \qquad (2.2)$$

Each generator must also observe minimum up and down time constraints; a generator may not be on for fewer than t_{up} consecutive stages or off for less than t_{dn} consecutive stages:

$$u_k(i) \geq I(1 \leq x_k(i) < t_{up}) \qquad (2.3)$$

$$u_k(i) \leq 1 - I(-t_{dn} < x_k(i) \leq -1) \qquad (2.4)$$

The power produced by each generator is constrained by load flow relations:

$$u_k(i)A_{Gi}P_{Gi} - \sum_{j=1}^{N_G} A_{TGiGj}P_{TGiGj} - \sum_{j=1}^{N_L} A_{TGiLj}P_{TGiLj} = 0 \qquad (2.5)$$

$$R_{Li}P_{Li} + \sum_{j=1}^{N_G} A_{TLiGj}P_{TLiGj} + \sum_{j=1}^{N_L} A_{TLiLj}P_{TLiLj} = 0 \qquad (2.6)$$

$$R_{Li}Q_{Li} + \sum_{j=1}^{N_G} A_{TLiGj}Q_{TLiGj} + \sum_{j=1}^{N_L} A_{TLiLj}Q_{TLiLj} = 0 \qquad (2.7)$$

The real and reactive power flow in a transmission line of conductance G_{ab} and susceptance B_{ab} are given by:

$$P_{Tab} = V_a^2 G_{ab} - V_a V_b G_{ab} \cos(\theta_a - \theta_b) - V_a V_b B_{ab} \sin(\theta_a - \theta_b) \tag{2.8}$$

$$Q_{Tab} = -V_a^2 B_{ab} + V_a V_b B_{ab} \cos(\theta_a - \theta_b) - V_a V_b G_{ab} \sin(\theta_a - \theta_b) \tag{2.9}$$

The load flow equations ((2.5) to (2.7)) are represented by a single vector function f:

$$\mathbf{f}(\mathbf{p}_G, \mathbf{p}_L, \mathbf{q}_L, \mathbf{a}_G, \mathbf{a}_T, \mathbf{r}_L) = 0 \tag{2.10}$$

The rationing of loads is determined by a predetermined rationing function $\mathbf{g}$:

$$\mathbf{r}_L = \mathbf{g}(\mathbf{a}_G, \mathbf{a}_T, \mathbf{p}_G, \mathbf{u}_k, \mathbf{x}_k, \mathbf{p}_L, \mathbf{q}_L) \tag{2.11}$$

The vectors in these last two equations are defined in Table 2.2. There are several operating limits that must be observed:

$$V_{Gi}^{min} \leq V_{Gi} \leq V_{Gi}^{max} \tag{2.12}$$

$$P_{Gi}^{min} \leq P_{Gi} \leq P_{Gi}^{max} \tag{2.13}$$

$$Q_{Gi}^{min} \leq Q_{Gi} \leq Q_{Gi}^{max} \tag{2.14}$$

Within the load flow constraints and operating limits, the generation levels are chosen according to optimal power flow:

$$
\begin{aligned}
\mathbf{p}_G = \arg\min_{\mathbf{p}_G} \Bigg[&\sum_{i=1}^{N_G} c_i(P_{Gi}) \\
&+ \sum_{i=1}^{N_G} \sum_{j=1}^{N_G} I(|P_{TGiGj}| > P_{TGiGj}^{max}) A_{PGiGj} (P_{TGiGj} - P_{TGiGj}^{max})^2 \\
&+ \sum_{i=1}^{N_G} \sum_{j=1}^{N_L} I(|P_{TGiLj}| > P_{TGiLj}^{max}) A_{PGiLj} (P_{TGiLj} - P_{TGiLj}^{max})^2 \\
&+ \sum_{i=1}^{N_L} \sum_{j=1}^{N_G} I(|P_{TLiGj}| > P_{TLiGj}^{max}) A_{PLiGj} (P_{TLiGj} - P_{TLiGj}^{max})^2 \\
&+ \sum_{i=1}^{N_L} \sum_{j=1}^{N_L} I(|P_{TLiLj}| > P_{TLiLj}^{max}) A_{PLiLj} (P_{TLiLj} - P_{TLiLj}^{max})^2 \Bigg]
\end{aligned} \tag{2.15}
$$

As before, the indicator variables have a value of 1 if true and 0 if false. The four double summations in the optimization are "soft" constraints on line flow congestion; for each line flow that exceeds the maximum limit P_{Tab}^{max}, a quadratic penalty function is added to the total cost.

t	Time of day (hours)
h_k	Number of hours in stage k
N_G	Number of generators
N_L	Number of loads
P_{Gi}	Real power from generator i
Q_{Gi}	Reactive power at generator i
P_{Li}	Real power at load i
Q_{Li}	Reactive power at load i
P_{Tab}	Real power flow in transmission line from a to b
Q_{Tab}	Reactive power flow in transmission line from a to b
V_{Gi}	Voltage magnitude at generator i
θ_{Gi}	Voltage angle at generator i
V_{Li}	Voltage magnitude at load i
θ_{Li}	Voltage angle at load i
A_{Gi}	Availability of generator i (1 = Available, 0 = Not available)
A_{Tab}	Availability of transmission line between a and b
A_{Pab}	Penalty cost factor for transmission line between a and b
$c_{Gi}(P_{Gi})$	Cost of generation for generator i
R_{Li}	Rationing of load i (1 = Served, 0 = Dropped)
I_{Li}	Insurance payment to load i in event of loss of service
S_i	Startup cost for generator i
T_i	Shutdown cost for generator i
p_i	Unit price for load i ($/kWh)
ρ_i	Reliability of service for load i
f	Load flow equations
g	Rationing function (known a priori)
$u_k(i)$	Control decision at stage k for generator i (1 = On, 0 = Off)
$x_k(i)$	State at stage k of generator i

Table 2.1: Quantities included in the cost equation for unit commitment.

$\mathbf{p}_G = [P_{G1}\, P_{G2} \ldots P_{GN_G}]^T$
$\mathbf{p}_L = [P_{L1}\, P_{L2} \ldots P_{LN_L}]^T$
$\mathbf{q}_L = [Q_{L1}\, Q_{L2} \ldots Q_{LN_L}]^T$
$\mathbf{a}_G = [A_{G1}\, A_{G2} \ldots A_{GN_G}]^T$
$\mathbf{a}_T = [A_{TG1G2} \ldots A_{TGN_GLN_L}\, A_{TL1G1} \ldots A_{TL(N_L-1)LN_L}]^T$
$\mathbf{r}_L = [R_{L1}\, R_{L2} \ldots R_{LN_L}]^T$
$\mathbf{u}_k = [u_k(1)\, u_k(2) \ldots u_k(N_G)]^T$
$\mathbf{x}_k = [x_k(1)\, x_k(2) \ldots x_k(N_G)]^T$

Table 2.2: Vector quantities used in the load flow and optimal power flow equations.

2.2   Present Unit Commitment Solution Methods

A large variety of solution techniques for unit commitment scheduling have been implemented and proposed [15]. These techniques include priority list methods [15], Lagrangian relaxation with subproblems solved by dynamic programming [2, 3, 16], and dynamic programming with branch and bound search filtering [4]. In these techniques, a finite cost horizon is used, ranging from eight hours to a week or more [4, 15, 16].

The possibility of generator failures is typically handled by providing an adequate reserve margin [2, 3, 4]. Other schemes are available for a more detailed reliability analysis. One scheme uses Monte Carlo simulation of generator outage scenarios [17]; the possible scenarios can also be enumerated [17]. Monte Carlo simulation is used in [18, 19] to analyze future load patterns for medium term planning (about 5 years).

One key aspect of the unit commitment formulation in this chapter is the stochastic nature of the problem. Many methods in use are deterministic with respect to the load power, meaning that the optimization is performed assuming that the demand is equal to the forecasted value [3, 4]. This method produces a certainty equivalence controller (CEC) [1], since a random variable has been replaced by its expected value. Some methods do allow for a probabilistic distribution of the load power [16, 20]. Also, many formulations simply require that total generation exceed total demand and ignore some details of the network, such as losses [4, 16].

2.3   Interruptible Service Contracts

At present, reserve requirements are based on the $(N-1)$ criterion, which means that there must be sufficient reserve on the system such that no load will lose power if any one line or any one generator fails [21]. Our formulation allows for the possibility of customer choice of interruptible service for a reduced rate. In this scenario, a customer chooses service with a given reliability ρ for a given price p [22]. A discrete number of contracts are available, including an option for maximum reliability. If all customers choose the maximum reliability, the problem will be the same as the current $(N-1)$ criterion; otherwise, the utility will be allowed to drop some loads in the event of a component failure. The rationing of a load is associated with a contingency that is at least as severe as a minimum contingency level specified in the service contract. To compensate the customer for loss of service, it is assumed that the utility makes an insurance payment I to the customer for loss of service, for which the customer regularly pays a premium $(1-\rho)I$ [22].

The formulation of reliability levels in [22] presumes that the total supply available takes on discrete values with known probabilities. Given a set of generators with maximum generating limits and failure probabilities, a set of contracts with known reliabilities can be obtained. One method is to consider every possible combination of at most l_f generator failures. The probabilities of each such combination may be used to devise a set of contracts.

CHAPTER 3

UNIT COMMITMENT IN A DEREGULATED ENVIRONMENT

So far, the focus has been on unit commitment in the current regulated environment; however, the approach is easily extended to the deregulated environment of the future. In one possible scenario, the unit commitment problem becomes one of dispatching units solely to meet reserve requirements, while base load is provided by other generators through bilateral contracts. Another possibility involves an Independent System Operator (ISO) receiving bids for available generation from various individual power producers after indicating at what price power will be bought and sold. The ISO then uses the given unit commitment formulation to determine which units will actually be used. Generally, individual owners of generation will determine a unit commitment strategy only for the generation units that they own; this book will focus on answering the question of how individual power producers can make optimal unit commitment decisions.

3.1   Possible Formats for the Electricity Market

At the time of this writing, discussion on the configuration of the market-place in a deregulated environment is centering on two basic models. The first model is referred to as a "poolco" setup [10]. In this format, a single entity, generally called an Independent System Operator (ISO), acts as a middleman for all power transactions; i.e., all customers buy their power from the ISO, and all of the individual power producers sell their power to the ISO. The second type of marketplace being considered is known as bilateral contracts [11, 23]. In this scenario, power customers can buy directly from specific generators at prices and terms of mutual agreement. An ISO exists; however, its function in a world of bilateral contracts is to determine whether the complete set of bilateral contracts is feasible on the transmission grid. A feasible set of transactions is one that does not violate any load flow restrictions (transmission line flows, voltage restrictions, etc.) or dynamic stability constraints. The ISO also provides ancillary services such as frequency regulation. Many models currently under discussion are a hybrid scheme, consisting of a power pool, but also allowing the arrangement of bilateral contracts between large buyers and sellers [24, 25].

3.1.1   The Poolco Marketplace

Currently, the electricity market is projected to be an hourly market, with the bidding taking place a day in advance. The ISO in a poolco market is obligated to find the price which balances supply and demand. This price may be found in different ways. In one method, each supplier and buyer must submit as their bid a price schedule; the amount that the participant is willing to buy or sell at many different prices. The ISO then takes the most expensive price bid submitted for supply that is needed to satisfy demand and the market clearing price is set equal to that bid. All generators are paid at the market clearing price for their bids [5]. An alternative is to have the ISO begin by guessing a price and receiving bids for generation and demand at that price. The ISO then adjusts the price to offset the imbalance and receives new offers for supply and demand. The process iterates until convergence at an equilibrium is reached.

3.1.2   The Market of Bilateral Contracts

Bilateral contracts can have essentially any length from hours to months to even years. Throughout this book, the term "bilateral market" refers to a competitive spot market of hour-long bilateral contracts. Longer term bilateral contracts are identical to a series of hour-long forward contracts over the term of the long-term contract; forward contracts are discussed in detail in Chapter 8. In a bilateral marketplace, it is generally assumed that the market forces of supply and demand will result in the price converging to the equilibrium, just as it does in other markets. Bilateral contracts may also be obtained by loads or groups of loads with generators for reserve to provide a back-up supply if the original supplier should experience a failure.

3.2   Unit Commitment for an Individual Supplier

We will now examine how the owner of a single generator makes optimal unit commitment decisions under the marketplace formats described in the previous section. Note that the objective of unit commitment under deregulation is to maximize profit, while under the regulated industry formulation of Chapter 2, the goal is to minimize cost.

3.2.1 Bilateral Market

From the perspective of an individual power producer, this case is simpler; therefore, we first examine unit commitment in a competitive bilateral market. For this formulation, we assume that the generator is capable of selling as much power as desired at the market equilibrium price p_k for hour k. The only control for the problem is the zero-one variable u_k, the decision whether to turn on or off at hour k. The generation level, P_G, may be regarded as a function of the control u_k and the price p_k. If $u_k = 0$, then $P_G = 0$. If $u_k = 1$, then P_G at hour k is set to maximize the profit:

$$p_k P_G - c_G(P_G)$$

For a quadratic cost function:

$$c_G(P_G) = aP_G^2 + bP_G + c \tag{3.1}$$

the derivative of profit with respect to P_G is:

$$p_k - 2aP_G - b$$

Setting this derivative to zero, we find the value of P_G which maximizes profits for hour k:

$$P_G = \frac{p_k - b}{2a} \tag{3.2}$$

The total profit π_k for the producer in hour k is:

$$\pi_k = u_k(p_k P_G - c_G(P_G) - I(x_k < 0)S) - (1 - u_k)(c_f + I(x_k > 0)T) \tag{3.3}$$

c_f are fixed costs incurred during an hour where the generator is off; it is implied that $c > c_f$, meaning that the constant term of the quadratic cost function includes these fixed costs. The remaining symbols have the same meaning as in Chapter 2. The price p_k is, from the supplier's perspective, an exogenous random variable with some probability distribution. Notice that this distribution reflects the uncertainty in demand. Modeling of the price is considered in detail in Chapter 6.

Note that in comparison to Chapter 2, there is only one control variable. The unit commitment problem under deregulation is much simpler to solve than the corresponding problem in the current regulated environment since the number of controls is greatly reduced. This simplification also makes it possible to consider more details in unit commitment; in particular, a stochastic problem formulation becomes feasible to solve.

3.2.2   Poolco Market

In a poolco market, the general unit commitment formulation for individual producers is complicated and raises several thorny issues; however, with several simplifying assumptions, the problem may be solved in a similar manner to the bilateral market. In general, the profit for a single independently owned generator during one hour is:

$$\pi_k = u_k(p_k P_G - c_G(P_G) - I(x_k < 0)S) - (1 - u_k)(c_f + I(x_k > 0)T) \qquad (3.4)$$

$$P_G = A_\% P_B \qquad\qquad\qquad\qquad\qquad\qquad\qquad\qquad (3.5)$$

p_k is the price of power, P_B is the amount of power that the supplier offers to sell at price p_k during hour k, and $A_\% \in [0, 1]$ is the percentage of the bid that was accepted by the ISO.

There are two ways that the ISO can respond to bids. The first is a simple yes/no decision; either the bid is accepted in full or rejected. For this case, $A_\%$ only takes the discrete values 0 and 1. The second possibility is that the ISO may also be allowed to partially accept bids; the generator is permitted to sell a percentage of the original bid amount. In this case, the owner would undoubtedly set a minimum acceptance percentage; if the ISO is unwilling to allow the generator to sell at least this minimum percentage of the bid, then the bid is not accepted. This minimum is created by two conditions; first, the generator itself has a physical minimum generation constraint (P_G^{min}), and second, because of fixed costs of generation as well as startup costs, producing at a low generation level may cause the owner to incur a loss. Because of these conditions, the owner would rather not produce any power than generate a small amount. In this type of a marketplace, $A_\%$ is continuously distributed between $A_\%^{min}$ and 1 but can take the values 0 and 1 with positive probability. If bids are only accepted in full or rejected, then the controls are u_k, the decision to turn on or off the generator at each hour, and P_B, the bid amount at each hour. If the marketplace format allows partial acceptance of bids, then there are three control variables: u_k, P_B, and $A_\%^{min}$.

From the supplier's perspective, p_k and $A_\%$ are random variables; however, estimating the distribution of $A_\%$ may be difficult, especially since $A_\%$ is a function of the bid schedule. $A_\%$ is also influenced by gaming issues; a supplier may simply bid at zero price in order to be assured of getting scheduled, and hope (or assume) that a high-cost generator at the margin will set a profitable market-clearing price for all participants.

A problem of interest for this type of market structure is the situation where a bid from an individual generator is not accepted for a given period, but is accepted for the periods immediately before and after the rejected period. It is unclear what the producer's options are in this case [25]. Another topic of concern in pool-based markets is the observed flatness of the optimum for unit commitment in large systems. This characteristic of unit commitment means that the ISO can choose among many unit commitment strategies that have virtually identical total costs; however, the profits of individual producers vary widely among the different unit commitment solutions [24].

Because of these complications and a lack of suitable models for their representation, the poolco unit commitment problem is largely outside the scope of this book. However, if the producer can reasonably assume $A_\% = 1$ for all bids, then the bilateral market formulation may be used, with modifications. If the generator is not able to choose generation level in response to price, then the amount of power bid will generally be chosen to maximize the expected profit:

$$p_k P_B - c_G(P_B)$$

Since p_k is the only random variable, the optimal bid amount is, for a quadratic cost function:

$$P_B = \frac{E\{p_k\} - b}{2a} \tag{3.6}$$

The profit for hour k is:

$$\pi_k = u_k(p_k P_B - c_G(P_B) - I(x_k < 0)S) - (1 - u_k)(c_f + I(x_k > 0)T) \tag{3.7}$$

3.3 Multiple Generation

As observed in Chapter 7, the complexity of the unit commitment problem under the present regulatory environment increases exponentially with the number of generators. However, in an idealized electricity marketplace under deregulation, an owner of several generators can optimize each generator individually, which means that the computation increases only linearly. The key difference between the two situations is that a utility in the present coordinated environment dispatches generation according to optimal power flow; if a given generator is turned off, then the output of all other generators will need to be readjusted in order to remain optimal. However, if the marketplace is competitive and no congestion constraints are present, then a generator owner who shuts down one generator will have negligible impact on the aggregate supply curve of the market, and hence the price will remain essentially unchanged. Since the optimal power setting of any generator in a competitive marketplace is determined by the price equals marginal cost rule, the profits of the owner's other generators will be unaffected. Of course, if the owner has significant market share, then this reasoning is no longer valid.

3.4   Secondary Market for Reliability

In a deregulated environment, an alternative method for selling variable reliability levels which parallels interruptible contracts is the creation of a secondary market for back-up power. In this scenario, the seller of power would arrange to buy reserve energy from another generator or generators to cover sales in the event that the original generator experiences a failure. The seller could even arrange an interruptible contract with one or more loads as an equivalent to purchasing reserve from other generators. In this formulation, the original contract between the buyer and seller would specify a level of reliability; the seller will then purchase reserve to meet the probability of service that was agreed upon in the contract. The reliability level would likely be specified as a penalty payment for loss of service; this payment implicitly defines a probability of service. This idea is explored in detail in Chapter 9.

3.5   Other Issues

There are several other factors that can affect the results that are derived in this book. One such issue is market power. It is generally assumed that the generation marketplace is competitive. However, it is possible that some generator owners will have a significant market share, resulting in an oligopolistic market. Gaming theory is generally required to analyze these situations.

A second issue worth mentioning is stranded cost recovery for existing utilities. Many utilities have made investments in generation which will become obsolete and of little value in an open access market. Various methods have been proposed to compensate these utilities for these losses that are due to deregulation. The planning horizon for the problems of interest in this book are short enough such that capital is regarded as a fixed cost. The producers are assumed to make capital investments over a much longer time horizon in anticipation of recovering the capital costs through future operating surpluses.

Another issue is the pricing of power under congested transmission line conditions. There is a large debate over what method should be used for the pricing of power in these situations. In general, transmission congestion is handled in this book by assuming that the price for power sold reflects congestion if it exists. Congestion is examined in more detail in Chapter 10.

CHAPTER 4
SURVEY OF THE DYNAMIC PROGRAMMING FORMULATION

Mathematically, commitment decision problems can be expressed as dynamic programming (DP) problems, including control inputs, system states, and uncertain (random) quantities. Time is broken down into a series of stages, and a control decision is made at the beginning of each stage. The system can be described by the following equations [1]:

$$\mathbf{x}_{k+1} = \mathbf{f}_k(\mathbf{x}_k, \mathbf{u}_k, \mathbf{w}_k) \tag{4.1}$$

where $k = 0, 1, \ldots$ is the time index, $\mathbf{x}_k$ is the state vector at time k, $\mathbf{u}_k$ is the control input at time k, and $\mathbf{w}_k$ is a random disturbance. The control $\mathbf{u}_k$ is constrained to be in the set of admissible controls $U_k(\mathbf{x}_k)$ and is usually chosen by:

$$\mathbf{u}_k = \mu_k(\mathbf{x}_k) \tag{4.2}$$

A set of functions $\mu_k(\mathbf{x}_k)$ for all k is defined as a control policy.

At each stage, there is a cost to be paid. This cost may be negative, meaning that a reward is received. The problem is to determine a control policy that minimizes the cost (or maximizes the reward). The exact definition of minimal cost depends on whether the planning horizon is finite or infinite.

4.1 Finite Horizon Problems

A finite horizon means that the total cost over a specified number of stages is to be minimized. The number of stages is denoted by N. At each stage

k, a cost $g_k(\mathbf{x}_k, \mathbf{u}_k, \mathbf{w}_k)$ is incurred. Additionally, there is a terminal cost $g_N(\mathbf{x}_N)$ which depends on the final value of the state vector. The object of the problem is to find the control policy that minimizes the total expected cost over N stages; this is known as the optimal policy. Dynamic programming is an algorithm to find the optimal policy; the algorithm is expressed mathematically as [1]:

$$J_N(\mathbf{x}_N) = g_N(\mathbf{x}_N) \tag{4.3}$$

$$J_k(\mathbf{x}_k) = \min_{\mathbf{u}_k \in U_k(\mathbf{x}_k)} \underset{\mathbf{w}_k}{E} \left\{ g_k(\mathbf{x}_k, \mathbf{u}_k, \mathbf{w}_k) + J_{k+1}(\mathbf{f}_k(\mathbf{x}_k, \mathbf{u}_k, \mathbf{w}_k)) \right\} \tag{4.4}$$

where $\underset{\mathbf{w}_k}{E}$ denotes the expected value operator with respect to the random variables $\mathbf{w}_k$. $J_k(\mathbf{x}_k)$ denotes the optimal expected cost when beginning at stage k. The DP algorithm begins by finding the optimal cost-to-go for the last stage. The algorithm then iterates backwards in time to calculate $J_k(\mathbf{x}_k)$ having already calculated $J_{k+1}(\mathbf{x}_{k+1})$, and the iteration continues until stage 0 (the current stage) is reached. An optimal policy is obtained as a set of functions $\mu_k^*(\mathbf{x}_k)$ such that $\mathbf{u}_k^* = \mu_k^*(\mathbf{x}_k)$ attains the minimization in equation (4.4) for each $\mathbf{x}_k$ and k; note that the optimal policy need not be unique.

4.2   Infinite Horizon Problems with a Discount Factor

In many cases, it is desirable to minimize the total cost over a very large number of stages. In these situations, posing the problem as a finite horizon problem with a large N is impractical because of the enormous amount of computation involved. A simpler solution is to use an infinite cost horizon, meaning that the cost is minimized over an infinite number of stages. In order to make sure that the problem is well defined, several approaches are possible. In this section, we consider the case where the cost of the next stage is discounted by a factor α such that $0 < \alpha < 1$ and the cost per stage has a finite upper bound, so that the cost over the infinite horizon is finite.

We will assume that the state transition equation, cost per stage, and probability distribution of the random disturbances is the same for all stages. Notice that under these assumptions, the time subscript k on the state and control is superfluous, since being at time 0 and looking ahead to an infinite number of stages is identical to being at time 1 (or 100) and looking ahead to an infinite number of stages. Furthermore, the optimal policy is a stationary policy, meaning that it is the same function μ for all time steps.

To simplify the analysis, we will assume for the infinite horizon problems that there are a finite number of states, numbered from 1 to n. The transition probability from state i to state j with the application of control u is denoted by $p_{ij}(u)$. If we are currently at state i and control u is applied, then the expected cost for the current stage is denoted as $g(i, u)$.

Using the same line of reasoning as before, it is clear that the optimal cost-to-go function is only a function of the state and not of the time step. Therefore, the finite horizon DP equation (4.4) becomes the following equation, known as Bellman's Equation, for an infinite horizon problem with cost discounting [1]:

$$J^*(i) = \min_{u \in U(i)} \left[g(i, u) + \alpha \sum_{j=1}^{n} p_{ij}(u) J^*(j) \right] , \; i = 1, \ldots, n \qquad (4.5)$$

Bellman's Equation is actually a system of equations, and $J^*(i)$ represents the optimal total expected cost when starting at state i. The costs $J^*(i)$ for $i = 1, \ldots, n$ are the unique solution to Bellman's Equation [1], although this does not imply that there is a unique optimal policy.

4.3 Stochastic Shortest Path Problems

A second type of infinite horizon problem is one that has no discounting factor but does have a termination state, denoted as t. Once the system enters the termination state, it remains in that state and no more costs are incurred. This is expressed mathematically as $p_{tt}(u) = 1$ and $g(t, u) = 0$. The problem is well defined if for all admissible policies there is positive probability that the system will enter the termination state in no more than m stages, for some integer m [1]. For the stochastic shortest path problem, Bellman's Equation is [1]:

$$J^*(i) = \min_{u \in U(i)} \left[g(i, u) + \sum_{j=1}^{n} p_{ij}(u) J^*(j) \right] , \; i = 1, \ldots, n \qquad (4.6)$$

It is worth noting that:

$$\sum_{j=1}^{n} p_{ij}(u) = 1 - p_{it}(u) \leq 1 \qquad (4.7)$$

since there may be a positive probability of entering the termination state from state i.

4.4   Average Cost per Stage Problems

A third type of infinite horizon problem is one that considers the average cost per stage, not the total cost over an infinite horizon. For a policy Π, the average cost per stage is defined as [1]:

$$J_\Pi(i) = \lim_{N \to \infty} \frac{1}{N} E \left\{ \sum_{k=0}^{N-1} g(x_k, \mu_k(x_k)) | x_0 = i \right\} \qquad (4.8)$$

Note that the average cost per stage of a policy is the same for initial states i and j if, when starting at state i, state j is eventually reached with probability 1, or vice versa. The average cost for the two initial states is identical because the expected time to reach j from i (or i from j) is finite, and therefore the costs incurred during the transition from i to j contribute nothing to the average cost per stage (the sum of a finite number of terms multiplied by $1/N$ approaches zero as $N \to \infty$). Similarly, if any state can be reached from any other state with probability 1 by any policy, then $J^*(i) = J^*(j)$ for any i and j between 1 and n [1].

Since in most average cost problems the optimal average cost is independent of the state, a stochastic shortest path problem which is equivalent to the original average cost problem can be used for analysis. One state, henceforth assumed to be state n, is assumed to be visited with positive probability during the first m stages, regardless of the policy used, where m is some positive integer. The stochastic shortest path problem with an expected cost per stage of $g(i, u) - \lambda^*$ for state i, and in which any transition to state n is replaced by a transition to a termination state t, is an equivalent problem to the original average cost problem. λ^* is the optimal average cost per stage.

The reasoning for the equivalence of the two problems may be summarized by noting that the average cost per stage for a stationary policy μ may be written as:

$$\lambda_\mu = \frac{C_{nn}(\mu)}{N_{nn}(\mu)} \qquad (4.9)$$

where $C_{nn}(\mu)$ is the expected cost from starting at state n until the next return to n and $N_{nn}(\mu)$ is the expected number of stages to return to n when starting at state n. Since $\lambda_\mu \geq \lambda^*$:

$$C_{nn}(\mu) - N_{nn}(\mu)\lambda^* \geq 0 \qquad (4.10)$$

Furthermore, the expected cost of the average cost problem under the optimal policy μ^* when starting at state n is equal to:

$$C_{nn}(\mu^*) - N_{nn}(\mu^*)\lambda^* = 0 \qquad (4.11)$$

The expected cost of the associated stochastic shortest path problem is denoted by $h^*(i)$ for the initial state i. Notice from equation (4.11) that $h^*(n) = 0$. Bellman's Equation for the average cost problem, which is identical to Bellman's Equation for the associated stochastic shortest path problem, may be written as [1]:

$$\lambda^* + h^*(i) = \min_{u \in U(i)} \left[g(i, u) + \sum_{j=1}^{n} p_{ij}(u) h^*(j) \right], \quad i = 1, \ldots, n \qquad (4.12)$$

$$h^*(n) = 0$$

Like the other infinite horizon problems, this form of Bellman's Equation has a unique solution. The preceding argument and a formal proof are found in [1].

CHAPTER 5

UNIT COMMITMENT FOR AN INDIVIDUAL POWER PRODUCER

As mentioned earlier, the unit commitment problem is used throughout this book to illustrate the characteristics of commitment decision problems. In this chapter, we examine unit commitment scheduling in a scenario where all power producers decide individually whether to operate their generators or not during a given time period. In this formulation, the producer makes a unit commitment decision before deciding how much power to sell in the market and setting the generation levels on each generator. At the time of each unit commitment decision, the first hour's price may either be a known value or a random variable of some mean and variance. Throughout the remainder of this book, this price will generally be treated as unknown.

The unit commitment problem fits nicely into the dynamic programming framework, in which each stage of the problem is a one hour time interval. The problem to be solved by the generator owner is to maximize profit over a long term period. This problem may be expressed in three ways: 1. A finite horizon problem with a fairly high number of periods, 2. An infinite horizon problem of minimization of the average total cost, or 3. An infinite horizon problem with a discount factor reflecting inflation and interest rates. The examples in this book will use a 24-hour horizon; formulating an infinite horizon is difficult because the expected price is a periodic function of the time of day. In the dynamic programming framework, the state is x_k, the generator status, while the control is the on/off decision, denoted u_k. Note that $u_k \in \{0, 1\}$, since there are only two possible decisions. Price is the only random disturbance input.

5.1   Unit Commitment without Generation Limits

We begin with the simplest formulation of the deregulated unit commitment problem; one in which there are no generation limits, and the producer receives the same price for all power sold during a single market time period. We further assume that the price p_k for each time period k is a random variable with mean $\bar{p}_k$ and standard deviation σ_{pk}, which are estimated from past data.

The only control for the problem is u_k, the decision whether to turn on or off at stage k. The generation level, P_G, may be regarded as a function of the control u_k and the price p_k. If $u_k = 0$, then $P_G = 0$. If $u_k = 1$, then P_G at stage k is set to maximize the profit:

$$p_k P_G - c_G(P_G)$$

For a quadratic cost function:

$$c_G(P_G) = aP_G^2 + bP_G + c \tag{5.1}$$

the derivative of profit with respect to P_G is:

$$p_k - 2aP_G - b$$

Setting this derivative to zero, we find the value of P_G which maximizes profits for stage k:

$$P_G = \frac{p_k - b}{2a} \tag{5.2}$$

The profit π_k for the producer in stage k is:

$$\pi_k = u_k(p_k P_G - c_G(P_G) - I(x_k < 0)S) - (1 - u_k)(c_f + I(x_k > 0)T) \tag{5.3}$$

c_f are fixed costs incurred during a stage where the generator is off; it is implied that $c > c_f$, meaning that the constant term of the quadratic cost function includes these fixed costs. The profit per stage $\pi_k(x_k, u_k, p_k)$ is a function of the state, control, and the random disturbance; here and throughout this book, the notation of functional dependence is suppressed. Since π_k depends on the random variable p_k, it has an expected value:

$$
\begin{aligned}
\operatorname*{E}_{p_k}\{\pi_k\} \;=\; & u_k(\operatorname*{E}_{p_k}\{p_k P_G - c_G(P_G)\} - I(x_k < 0)S) \\
& - (1 - u_k)(c_f + I(x_k > 0)T)
\end{aligned} \tag{5.4}
$$

Substituting for P_G, this becomes:

$$\mathop{E}_{p_k}\{\pi_k\} = u_k \left(\mathop{E}_{p_k} \left\{ \frac{p_k^2 - bp_k}{2a} - a\frac{(p_k - b)^2}{4a^2} - b\frac{p_k - b}{2a} - c \right\} - I(x_k < 0)S \right)$$
$$- (1 - u_k)(c_f + I(x_k > 0)T) \tag{5.5}$$

The last two fractions in the expectation may be combined to form:

$$\mathop{E}_{p_k}\{\pi_k\} = u_k \mathop{E}_{p_k} \left\{ \frac{p_k^2 - bp_k}{2a} - \frac{p_k^2 - 2p_k b + b^2 + 2p_k b - 2b^2}{4a} - c \right\}$$
$$- u_k I(x_k < 0)S - (1 - u_k)(c_f + I(x_k > 0)T) \tag{5.6}$$

which may be rewritten as:

$$\mathop{E}_{p_k}\{\pi_k\} = u_k \left(\mathop{E}_{p_k} \left\{ \frac{p_k^2 - bp_k}{2a} - \frac{p_k^2 - b^2}{4a} - c \right\} - I(x_k < 0)S \right)$$
$$- (1 - u_k)(c_f + I(x_k > 0)T) \tag{5.7}$$

Note (see equation (E.14)) that the expected value of the square of a random variable is the sum of the mean squared and the variance (standard deviation squared):

$$\mathop{E}_{p_k}\{p_k^2\} = \overline{p}_k^2 + \sigma_{pk}^2 \tag{5.8}$$

The expected profit for stage k then becomes:

$$\mathop{E}_{p_k}\{\pi_k\} = u_k \left(\frac{\overline{p}_k^2 - b\overline{p}_k + \sigma_{pk}^2}{2a} - \frac{\overline{p}_k^2 + \sigma_{pk}^2 - b^2}{4a} - c - I(x_k < 0)S \right)$$
$$- (1 - u_k)(c_f + I(x_k > 0)T) \tag{5.9}$$

Note that the fractions in equation (5.9) can be combined; we have left them separate to distinguish expected generation cost from expected revenue. Note that we did not initially assume any particular distribution for p_k in the derivation of equation (5.9); this equation will give the expected profit per stage for any distribution of the price.

5.2 Unit Commitment with Generation Limits

We now consider the effect that minimum and maximum generation limits have on the expected profit. Recall from equation (5.2) that power is a linear function of price; therefore, the power generated will be a random variable that has a distribution with the same shape as that of the price. When we include generation limits, the distribution of the power becomes a truncated distribution, with a finite probability that the power is at either the upper or lower limit. To calculate the expected profit per stage when generation limits are present, we will need to examine in detail the properties of a truncated probability distribution. Truncated versions of the normal and lognormal distributions are presented here; the latter case is particularly useful for price models (see Chapter 6).

5.2.1   Truncated Normal Distributions

Throughout this discussion, we will assume that X is a normally distributed random variable with mean m and standard deviation σ. Z is a truncated normal variable with a minimum Z_{min} and a maximum Z_{max}; Z may be generated from a normal random variable X by:

$$
Z = \begin{cases} Z_{min} & X \leq Z_{min} \\ X & Z_{min} < X < Z_{max} \\ Z_{max} & X \geq Z_{max} \end{cases}
\tag{5.10}
$$

Therefore, Z is continuously distributed between Z_{min} and Z_{max} but also has a finite probability of being equal to Z_{min} or Z_{max}. We are interested in finding the mean and variance of Z, as well as computing the expected value of XZ, as these are all quantities we will need when calculating the expected profit. The formulas for these quantities in terms of m and σ are presented here; the derivation of these equations is detailed in Appendix C.

The mean of Z is:

$$
E(Z) = m + L_{CF}
\tag{5.11}
$$

$$
\begin{aligned}
L_{CF} = {} & \frac{\sigma}{\sqrt{2\pi}} \left(e^{-\frac{(Z_{min}-m)^2}{2\sigma^2}} - e^{-\frac{(Z_{max}-m)^2}{2\sigma^2}} \right) + (Z_{min} - m)\left(\frac{1}{2}\right. \\
& + \frac{1}{2}\operatorname{erf}\left(\frac{Z_{min}-m}{\sigma\sqrt{2}}\right)\Big) + (Z_{max} - m)\left(\frac{1}{2} - \frac{1}{2}\operatorname{erf}\left(\frac{Z_{max}-m}{\sigma\sqrt{2}}\right)\right)
\end{aligned}
\tag{5.12}
$$

L_{CF} is a "correction factor" to account for the change in mean of Z due to the presence of upper and lower limits. The variance of Z may be calculated by:

$$
\begin{aligned}
\operatorname{var}(Z) = {} & (\sigma^2 + L_{CF}^2)\left(\frac{1}{2}\operatorname{erf}\left(\frac{Z_{max}-m}{\sigma\sqrt{2}}\right) - \frac{1}{2}\operatorname{erf}\left(\frac{Z_{min}-m}{\sigma\sqrt{2}}\right)\right) \\
& + (Z_{min} - m - L_{CF})^2\left(\frac{1}{2} + \frac{1}{2}\operatorname{erf}\left(\frac{Z_{min}-m}{\sigma\sqrt{2}}\right)\right) \\
& + (Z_{max} - m - L_{CF})^2\left(\frac{1}{2} - \frac{1}{2}\operatorname{erf}\left(\frac{Z_{max}-m}{\sigma\sqrt{2}}\right)\right)
\end{aligned}
$$

$$+ \frac{\sigma}{\sqrt{2\pi}} \left((Z_{min} - m - 2L_{CF}) e^{-\frac{(Z_{min}-m)^2}{2\sigma^2}} \right.$$

$$\left. - (Z_{max} - m - 2L_{CF}) e^{-\frac{(Z_{max}-m)^2}{2\sigma^2}} \right) \tag{5.13}$$

Finally, the quantity XZ, where Z is derived by truncation of X between Z_{min} and Z_{max}, has expected value:

$$
\begin{aligned}
E(XZ) \;=\; & m^2 + \sigma^2 + \frac{m\sigma}{\sqrt{2\pi}} \left(e^{-\frac{(Z_{min}-m)^2}{2\sigma^2}} - e^{-\frac{(Z_{max}-m)^2}{2\sigma^2}} \right) \\
& + (mZ_{min} - m^2 - \sigma^2)\left(\frac{1}{2} + \frac{1}{2}\operatorname{erf}\left(\frac{Z_{min}-m}{\sigma\sqrt{2}} \right) \right) \\
& + (mZ_{max} - m^2 - \sigma^2)\left(\frac{1}{2} - \frac{1}{2}\operatorname{erf}\left(\frac{Z_{max}-m}{\sigma\sqrt{2}} \right) \right) \tag{5.14}
\end{aligned}
$$

5.2.2 Truncated Lognormal Distributions

We will now consider the effects of limits on the mean and variance of a random variable whose logarithm is normally distributed. Given a truncated normal random variable Z with mean m, standard deviation σ, and limits Z_{min} and Z_{max}, we wish to find the mean and variance of e^Z. We also want to determine the expected value of $e^X e^Z$, where Z is a truncated version of the normally distributed variable X. As before, derivations of these results are in Appendix C.

The mean of a truncated lognormal variable may be written as:

$$E(e^Z) = C_{CF} m + L_{CF} \tag{5.15}$$

with the quantities C_{CF} and L_{CF} defined by:

$$C_{CF} = \frac{1}{2} e^{\frac{1}{2}\sigma^2} \left(\operatorname{erf}\left(\frac{Z_{max}-m-\sigma^2}{\sigma\sqrt{2}} \right) - \operatorname{erf}\left(\frac{Z_{min}-m-\sigma^2}{\sigma\sqrt{2}} \right) \right) \tag{5.16}$$

$$
\begin{aligned}
L_{CF} \;=\; & e^{Z_{min}} \left(\frac{1}{2} + \frac{1}{2}\operatorname{erf}\left(\frac{Z_{min}-m}{\sigma\sqrt{2}} \right) \right) \\
& + e^{Z_{max}} \left(\frac{1}{2} - \frac{1}{2}\operatorname{erf}\left(\frac{Z_{max}-m}{\sigma\sqrt{2}} \right) \right) \tag{5.17}
\end{aligned}
$$

As with the normal distribution, C_{CF} and L_{CF} are quantities that described how the mean of e^Z is affected by the limits on Z. The variance of a truncated lognormal random variable is:

$$\operatorname{var}(e^Z) = \frac{1}{2} e^{2m+2\sigma^2} \left(\operatorname{erf}\left(\frac{Z_{max}-m-2\sigma^2}{\sigma\sqrt{2}} \right) - \operatorname{erf}\left(\frac{Z_{min}-m-2\sigma^2}{\sigma\sqrt{2}} \right) \right)$$

$$+ e^{2Z_{min}} \left(\frac{1}{2} + \frac{1}{2} \operatorname{erf} \left(\frac{Z_{min} - m}{\sigma \sqrt{2}} \right) \right)$$

$$+ e^{2Z_{max}} \left(\frac{1}{2} - \frac{1}{2} \operatorname{erf} \left(\frac{Z_{max} - m}{\sigma \sqrt{2}} \right) \right) - (C_{CF} m + L_{CF})^2 \quad (5.18)$$

Finally, the expected value of $e^X e^Z$ is

$$E(e^{X+Z}) = \frac{1}{2} e^{2m + 2\sigma^2} \left(\operatorname{erf} \left(\frac{Z_{max} - m - 2\sigma^2}{\sigma \sqrt{2}} \right) - \operatorname{erf} \left(\frac{Z_{min} - m - 2\sigma^2}{\sigma \sqrt{2}} \right) \right)$$

$$+ \frac{1}{2} e^{Z_{min} + m + \frac{1}{2}\sigma^2} \left(1 + \operatorname{erf} \left(\frac{Z_{min} - m - \sigma^2}{\sigma \sqrt{2}} \right) \right)$$

$$+ \frac{1}{2} e^{Z_{max} + m + \frac{1}{2}\sigma^2} \left(1 - \operatorname{erf} \left(\frac{Z_{max} - m - \sigma^2}{\sigma \sqrt{2}} \right) \right) \quad (5.19)$$

5.2.3   Expected Profit of Generation

Having formulas for truncated normal variables, we can readily find the expected profit of generation in each stage. First, we define the upper and lower marginal cost limits as:

$$p_{MC}^{min} = 2a P_G^{min} + b \qquad (5.20)$$

$$p_{MC}^{max} = 2a P_G^{max} + b \qquad (5.21)$$

If $p_k < p_{MC}^{min}$, then the optimal amount of power to produce is P_G^{min}; similarly, if $p_k > p_{MC}^{max}$, then profit is maximized by selling P_G^{max}. This observation leads us to define $p_{MC(k)}$, the marginal cost of production, as a truncated random variable in terms of the price p_k:

$$p_{MC(k)} = \begin{cases} p_{MC}^{min} & p_k \leq p_{MC}^{min} \\ p_k & p_{MC}^{min} < p_k < p_{MC}^{max} \\ p_{MC}^{max} & p_k \geq p_{MC}^{max} \end{cases} \qquad (5.22)$$

The choice of P_G which maximizes profits for stage k may thus be written as a function of $p_{(MC)k}$ for the case where P_G is constrained between P_G^{min} and P_G^{max}:

$$P_G = \frac{p_{(MC)k} - b}{2a} \qquad (5.23)$$

The expected profit in stage k is the same as equation (5.4), which was derived for the case without generation limits:

$$\begin{aligned} E_{p_k}\{\pi_k\} &= u_k (E_{p_k}\{p_k P_G - c_G(P_G)\} - I(x_k < 0)S) \\ &\quad - (1 - u_k)(c_f + I(x_k > 0)T) \end{aligned} \qquad (5.24)$$

Substituting equation (5.23) for P_G:

$$\mathop{E}_{p_k}\{\pi_k\} = u_k \left(\mathop{E}_{p_k} \left\{ \frac{p_k P_{(MC)k} - bp_k}{2a} - a\frac{(P_{(MC)k} - b)^2}{4a^2} - b\frac{P_{(MC)k} - b}{2a} - c \right\} \right.$$
$$\left. - I(x_k < 0)S \right) - (1 - u_k)(c_f + I(x_k > 0)T) \qquad (5.25)$$

The last two fractions may be combined via the same procedure as in equations (5.6) and (5.7):

$$\mathop{E}_{p_k}\{\pi_k\} = u_k \left(\mathop{E}_{p_k} \left\{ \frac{p_k P_{(MC)k} - bp_k}{2a} - \frac{p_{(MC)k}^2 - b^2}{4a} - c \right\} - I(x_k < 0)S \right)$$
$$- (1 - u_k)(c_f + I(x_k > 0)T) \qquad (5.26)$$

At this point, if the price p_k could be expressed as an independent identically distributed random variable, then the unit commitment problem would be essentially solved [26]. However, the price at time k is strongly correlated with the price at time $k - 1$, and therefore, it is necessary to augment the state of the system to include the price. This state augmentation is necessary because a high price in the next hour also implies high prices (and profits) during subsequent hours.

In order to continue, it is necessary to develop a price process model that reflects reasonably accurately the price correlations between hours. This price model for p_k is the subject of Chapter 6.

CHAPTER 6
PRICE PROCESS OF ELECTRICITY

In general, representing a price over several time periods as a series of independent random variables is a poor model, as it ignores correlation in the uncertainties between time periods. A more conventional model is to use a price process [6]. Developing a price process for electricity, however, is quite difficult as there is very little empirical data available. Furthermore, the demand for electricity is not the same over all hours of the day. As shown in Figure 6.1, the price is generally lowest in the early morning hours and highest in the early evening. This relation between hour of day and price closely follows the load profile (shown in Figure 6.2). In this book, price models are developed to account for the change in average price as a function of hour of day in addition to standard price models which treat all time steps equally. Additionally, the variance of the price estimate for the next time step may be significantly reduced by incorporating additional variables which are correlated with price, such as day of week or temperature. To create a price model for electricity, hourly data from the Pennsylvania-Jersey-Maryland (PJM) market [27], which opened on April 1, 1997, will be used.

6.1   Price Process Models

Prices for most common commodities, such as copper and oil, are typically modeled using a stochastic process, in which the price at a future time is a random function of the current price. A common price process model is Brownian motion (also called a Wiener process) which is a continuous-time process given by [6]:

$$dx = \epsilon_t \sqrt{dt} \qquad (6.1)$$

Since electricity markets typically operate on hourly intervals, a discrete-time process may be more appropriate. Brownian motion is in fact the limit of a

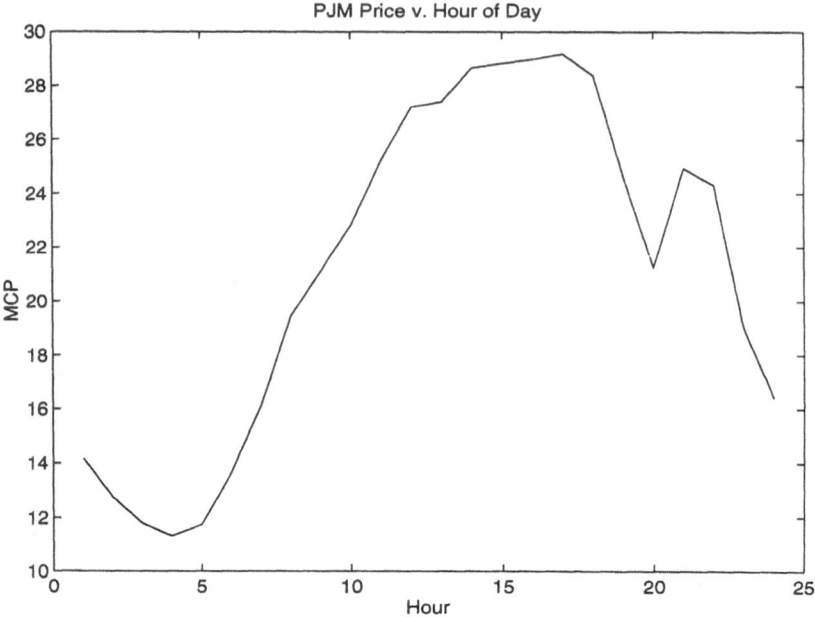

Figure 6.1: Average price vs. hour of day in the PJM electricity market.

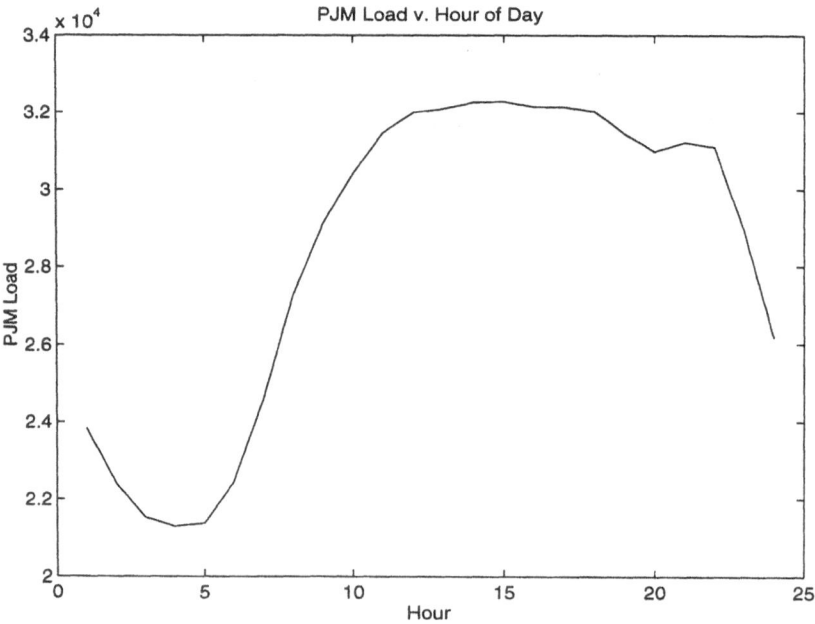

Figure 6.2: Average load vs. hour of day in the PJM electricity market.

discrete-time random walk:

$$x_t = x_{t-1} + e_t \tag{6.2}$$

as the time interval approaches zero [6]. e_t is normally distributed with zero mean and is independent of e_u for any $u \neq t$.

Another price process model is the mean-reverting process. Whereas a random walk can deviate far from the starting point and not return for a very long time, a mean-reverting process will have a tendency to return to a mean value over time. The mean-reverting process may be expressed in discrete time as the following:

$$x_t = \bar{x}(1 - e^{-\eta}) + e^{-\eta}x_{t-1} + e_t \tag{6.3}$$

which is known as the first order auto-regressive (AR(1)) process [6]. Here $\bar{x}$ is the mean and η is the rate of reversion; as $\eta \to 0$, the process becomes a random walk.

6.1.1 Long Term Model Behavior

Given the current price x_0, it is often important to ask what happens to the price many stages in the future. For the random walk model, the price at stage k may be written as:

$$x_k = x_0 + \sum_{i=1}^{k} e_i \tag{6.4}$$

Since e_i has zero mean, the expected value of x_k is simply:

$$E\{x_k\} = x_0 \tag{6.5}$$

Also, since the error terms are all independent of each other, the variance of x_k is:

$$\text{var}(x_k) = \sum_{i=1}^{k} \text{var}(e_i) \tag{6.6}$$

or, by defining $\sigma_e^2 = \text{var}(e_i)$:

$$\text{var}(x_k) = k\sigma_e^2 \tag{6.7}$$

Note that the variance tends to infinity as $k \to \infty$.

For a mean-reverting process, the equation for x_k is less obvious; however, it can be easily derived by induction:

$$x_k = \bar{x}(1 - e^{-k\eta}) + e^{-k\eta}x_0 + \sum_{i=1}^{k} e^{-(k-i)\eta}e_i \tag{6.8}$$

This relation may be verified by substituting into equation (6.3). Since e_i is assumed to be normally distributed, x_k will also be normally distributed [43]. The mean of x_k is:

$$E\{x_k\} = \bar{x}(1 - e^{-k\eta}) + e^{-k\eta}x_0 \tag{6.9}$$

The variance of x_k is calculated by applying the relation $\text{var}(cX) = c^2\text{var}(X)$ and again observing that the terms in the summation are all independent:

$$\text{var}(x_k) = \sum_{i=1}^{k} e^{-2(k-i)\eta}\text{var}(e_i) \tag{6.10}$$

By substituting $j = k - i$:

$$\text{var}(x_k) = \sum_{j=0}^{k-1} e^{-2j\eta}\sigma_e^2 \tag{6.11}$$

the summation is simply a geometric series. The formula for a geometric series is:

$$\sum_{i=0}^{k-1} \alpha^i = \frac{1 - \alpha^k}{1 - \alpha} \tag{6.12}$$

so the variance of x_k may be concisely written as:

$$\text{var}(x_k) = \sigma_e^2 \frac{1 - e^{-2k\eta}}{1 - e^{-2\eta}} \tag{6.13}$$

It is very interesting to observe that:

$$\lim_{k \to \infty} E\{x_k\} = \bar{x} \tag{6.14}$$

$$\lim_{k \to \infty} \text{var}(x_k) = \frac{\sigma_e^2}{1 - e^{-2\eta}} \tag{6.15}$$

Unlike the random walk, the variance of a mean-reverting process tends toward a constant for values in the distant future. It is important to note that this conclusion is inappropriate when using mean-reverting processes to model prices, as the parameters $\bar{x}$ and η will slowly change over time. However, for a relatively short-term horizon, equations (6.14) and (6.15) provide a good estimate of future prices.

6.1.2 Price Models using Logarithms

It should be noted that price models often are applied to the logarithm of the price rather than the actual price, as price changes are observed to be larger at higher price values (a lognormal distribution model), and the use of logarithms means that the price does not fall below zero [6]. Since the price models give the mean and variance for a logarithm of the price, it is necessary to convert these numbers to find the expected value and variance of the actual price. Given a normally distributed variable X with mean μ and variance σ^2, the expected value of e^X is:

$$E(e^X) = e^{\mu + \frac{1}{2}\sigma^2} \tag{6.16}$$

with variance:

$$\text{var}(e^X) = e^{2\mu + \sigma^2}(e^{\sigma^2} - 1) \tag{6.17}$$

These equations are derived in Appendix A.

6.1.3 Unit Root Tests

Given price data for a given commodity, it is possible to determine to a given degree of confidence whether or not the price process for the commodity is mean-reverting. This determination is made by using procedures known as unit root tests [28]. To test whether the price process is AR(1), the discrete-time price data is fitted to the regression model:

$$x_t = \alpha + \rho x_{t-1} + e_t \qquad 2 \le t \le n \tag{6.18}$$

where x_t is the price (or log of price) at time t. The test statistic ϕ_1 is computed from the regression as:

$$\phi_1 = \frac{(n-1)\hat{\sigma}_{NH}^2 - (n-3)\hat{\sigma}^2}{2\hat{\sigma}^2} \tag{6.19}$$

Here $\hat{\sigma}^2$ is the estimated variance from the regression model, and $\hat{\sigma}_{NH}^2$ is the estimated variance from the null hypothesis model (NH), in which the price follows a random walk:

$$x_t = x_{t-1} + e_t \tag{6.20}$$

Under NH, the estimate of variance is:

$$\hat{\sigma}_{NH}^2 = \frac{1}{n-1} \sum_{i=2}^{n} (x_i - x_{i-1})^2 \tag{6.21}$$

ϕ_1 is therefore simply the F-statistic of standard regression theory to test the alternate hypothesis of the mean-reverting model against the null hypothesis of a random walk [28, 29]. However, it turns out that if the null hypothesis is true, the test statistic ϕ_1 does not have an F-distribution as would be normally expected; instead, the value of ϕ_1 is checked against a set of empirically derived percentiles to determine whether the process is mean-reverting, to a given confidence level. For the same reason, a t-interval test for $\rho = 1$ using the regression estimate of the variance of $\hat{\rho}$ is not applicable [6, 28].

6.1.4   Application of Unit Root Test to PJM Electricity Data

The unit root test may be applied directly to the price data from the PJM electricity market. By performing the indicated regression on 502 hours of data (hours 2499 to 3000) using the natural logarithm of price, the parameter estimates are $\hat{\alpha} = 0.266$ and $\hat{\rho} = 0.917$, with an estimated variance of $\hat{\sigma}^2 = 8.95 \times 10^{-2}$. The value of the test statistic is $\phi_1 = 10.87$. For a random walk, the 99% level of ϕ_1 for a sample of $n = 500$ is 6.47 [28]. Since ϕ_1 for the PJM data is larger than 6.47, we can conclude that the PJM data is mean-reverting, with 99% confidence.

However, this model for the price treats all 24 hours of the day equally. Since the average load is higher at some hours than others, it would be expected that the average price for different hours would also vary. We will therefore examine whether the price process model can be improved by examining the load data.

6.2   Correlation of Price with Load

An examination of price data from the newly formed PJM market [27] reveals a clear correlation between the price of power and the system load. Figure 6.3 shows a plot of the market clearing price versus the PJM load for each hour in the data set. The price clearly is higher for larger load values, although there is significant variance in the price at any given load. The slope of price is moderate for intermediate load values but becomes very large for both low and high loads. A logarithmic plot of the price data (Figure 6.4) shows that the slope of the logarithm of price is relatively constant except for low load levels.

Applying linear regression to Figure 6.4 gives the following relation:

$$\ln P = \beta_1 L + \beta_0 \tag{6.22}$$

with the values for β_1 and β_0 given in Table 6.1. The estimated variance is $\hat{\sigma}^2 = 0.274$. (See section 6.4.1 for an overview of regression analysis.)

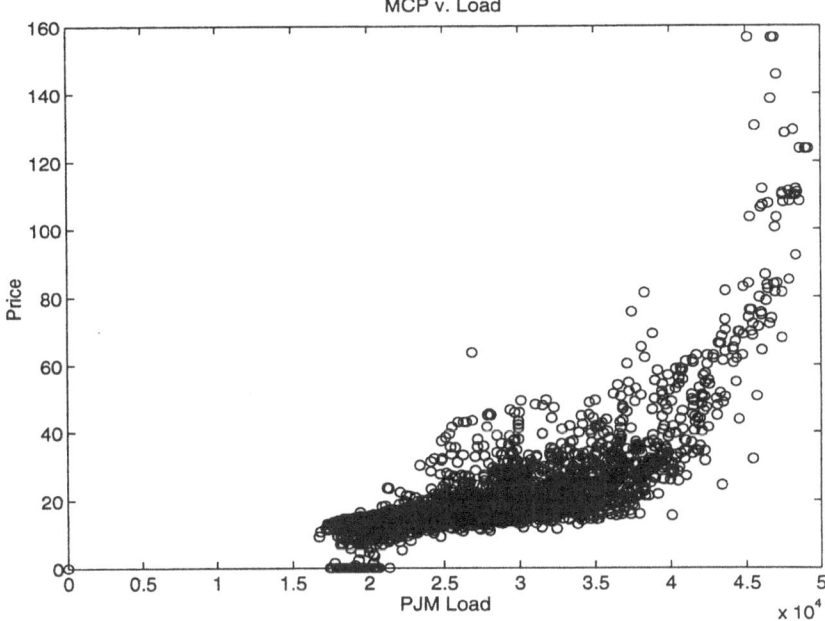

Figure 6.3: Scatter plot of price vs. load in the PJM electricity market.

Variable	Value	95% Confidence Interval	
		Lower	Upper
β_1	7.46×10^{-5}	7.17×10^{-5}	7.74×10^{-5}
β_0	7.55×10^{-1}	6.72×10^{-1}	8.39×10^{-1}

Table 6.1: Regression results for β_1 and β_0.

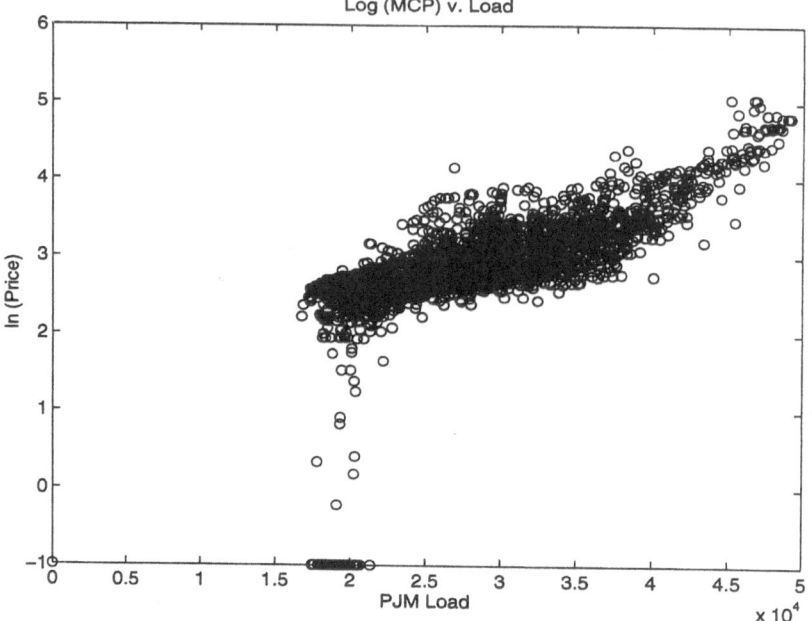

Figure 6.4: Scatter plot of natural logarithm of price vs. load in the PJM electricity market. Hours for which the market clearing price is zero are plotted as having a log of -1.

6.2.1 Expanded Price Process Models

In order to include load information in the price model, the following model is suggested:

$$x_t = x_{t-1} + m(L_t - L_{t-1}) + e_t \tag{6.23}$$

L_t is the actual load used (purchased) at time i. This model approximates the supply curve of the market as a line with a fixed slope and an intercept which follows a random walk. The demand curve is also approximated as a line of fixed slope. To see this, note that a linear supply curve relates price (P) to quantity (Q) by:

$$P = m_S Q + b_S \tag{6.24}$$

Similarly, a linear demand curve may be written as:

$$Q = L_0 - y_D P \tag{6.25}$$

where L_0 is the demand at a price of zero and y_D is the inverse of the slope. The equilibrium point of the market is the point at which the supply and demand curves cross:

$$L_0 - y_D P = \frac{P - b_S}{m_S} \tag{6.26}$$

The equilibrium price is given by:

$$P = L_0 \left(y_D + \frac{1}{m_S} \right)^{-1} + b_S (m_S y_D + 1)^{-1} \tag{6.27}$$

Let $L_{P'}$ denote the quantity demanded at price P':

$$L_{P'} = L_0 - y_D P' \tag{6.28}$$

The equilibrium price may then be written in terms of $L_{P'}$:

$$P = L_{P'} \left(y_D + \frac{1}{m_S} \right)^{-1} + y_D P' \left(y_D + \frac{1}{m_S} \right)^{-1} + b_S (m_S y_D + 1)^{-1} \tag{6.29}$$

From equation (6.29), the price is a linear function of the level of demand. $L_{P'}$ may be approximated by measuring the actual load. For the electricity market, which is observed to have very inelastic demand (y_D small), the error due to this approximation is small. This models treats the load as an exogenous random input of known mean and variance; L_t is an approximation of the demand curve at a given hour. Equation (6.23) is derived directly from:

$$x_t = m L_t + b_t \tag{6.30}$$

$$x_{t-1} = mL_{t-1} + b_{t-1} \tag{6.31}$$

which represent the price curve at times t and $t - 1$, and:

$$b_t = b_{t-1} + e_t \tag{6.32}$$

where b_t is the intercept of the price curve.

Not surprisingly, this model can be modified by treating the supply intercept as a mean-reverting process instead of a random walk. This modified model may be written mathematically as:

$$x_t = \bar{b}(1 - e^{-\eta}) + e^{-\eta}x_{t-1} + m(L_t - e^{-\eta}L_{t-1}) + e_t \tag{6.33}$$

This equation may be derived from the supply curve model of equations (6.30) and (6.31) and the equation for a mean-reverting process:

$$b_t = \bar{b}(1 - e^{-\eta}) + e^{-\eta}b_{t-1} + e_t \tag{6.34}$$

The parameters of the mean-reverting intercept model may be estimated using nonlinear regression, as described in Appendix A.

6.2.2   Test for Mean Reversion

We will now examine the price data from the PJM market to determine which of the above two models is more appropriate. A modification of the unit root test can be applied to answer this question. Specifically, we will calculate the F-statistic for the null hypothesis of equation (6.23) versus the alternate hypothesis of equation (6.33). This modified test statistic, which is denoted ϕ_{1m}, can then be compared against an estimate of its percentiles under the null hypothesis; these percentiles can be generated by Monte Carlo, similar to the method of [28].

The F-statistic for testing two hypotheses may be written:

$$F_{test} = \frac{\hat{\sigma}^2{}_{NH}d_{NH} - \hat{\sigma}^2 d}{\hat{\sigma}^2(d_{NH} - d)} \tag{6.35}$$

where, as before, $\hat{\sigma}^2{}_{NH}$ and $\hat{\sigma}^2$ are the estimates of variance under respectively the null and alternate hypotheses, and d_{NH} and d are the degrees of freedom in the null hypothesis and alternate hypothesis models. For n hours of price data encompassing $n - 1$ intervals, the test statistic ϕ_{1m} may be written as:

$$\phi_{1m} = \frac{(n - 2)\hat{\sigma}_{NH}^2 - (n - 4)\hat{\sigma}^2}{2\hat{\sigma}^2} \tag{6.36}$$

For the PJM data from hour 2499 to hour 3000, $\phi_{1m} = 36.85$. Based on 1000 Monte Carlo simulations of the random walk intercept model using the load values from the PJM data over the same 502 hour period, ϕ_{1m} is less than 6.02 with probability 0.99 under the null hypothesis model, which strongly suggests rejecting the model with the intercept as a random walk in favor of a mean-reverting intercept.

6.2.3   Price Prediction Algorithms

Now that we have several plausible price models, we will test each of them to see which one best predicts future prices. As mentioned earlier, it is often desirable to apply price process models to the logarithms of the price. From equation (6.16), the expected price should therefore be multiplied by $e^{\frac{1}{2}\sigma^2}$ to account for the bias resulting from taking an exponential of a normally distributed variable. However, a plot of the price over time, as shown in Figures 6.5 and 6.6, indicates that the variance estimate may be substantially different over different time ranges. Furthermore, the hours with zero and very low prices fall outside the line that describes the remainder of the data (as shown in Figure 6.4), and should probably not be included when estimating variance, as these points will cause the variance to be significantly overstated for most hours, leading to an exponential bias factor which is too large. The regression data for each model is given for three ranges: the complete range of 3096 hours, from hour 125 to hour 1225 (1100 intervals), and from hour 2649 to hour 3096 (447 intervals). The last two ranges do not include any hours with zero prices. Note also that the second interval has significantly less variance than the third interval; this may be a function of market startup, or possibly a seasonal volatility increase. The bias factor for each price model in the subsequent sections is estimated from the third range (hours 2649 to 3096). The task of estimating the variance at each hour from weighted least squares regression and devising an algorithm to discard outliers is left for future research.

Random walk

If the price process is modeled as a random walk, the estimate of the variance from the PJM price data is 31.84. The standard deviation of the error when predicting prices using a random walk model is therefore about 5.64. If the logarithm of price is modeled as a random walk, the variance estimates of $\ln P$ are given in Table 6.2. The exponential bias factor for this model is estimated as 1.021. However, in this case, the random walk model using logarithms results in a somewhat worse performance, with a prediction error having variance 32.79 and standard deviation of about 5.73.

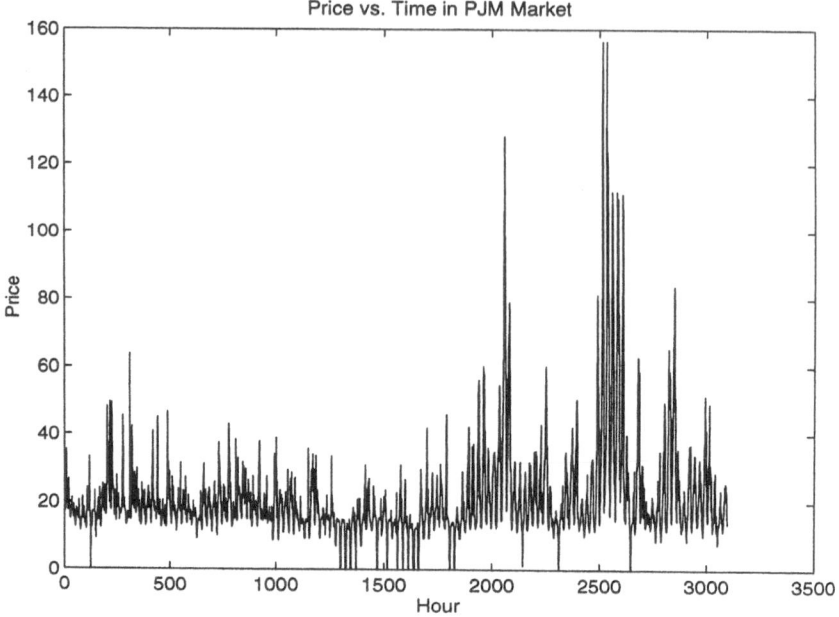

Figure 6.5: Plot of price vs. time in the PJM electricity market.

Hours	Degrees of Freedom	$\hat{\sigma}^2$
1 – 3096	3095	1.27×10^{-1}
125 – 1225	1100	2.80×10^{-2}
2649 – 3096	447	4.17×10^{-2}

Table 6.2: Variance estimates for random walk model of price logarithm.

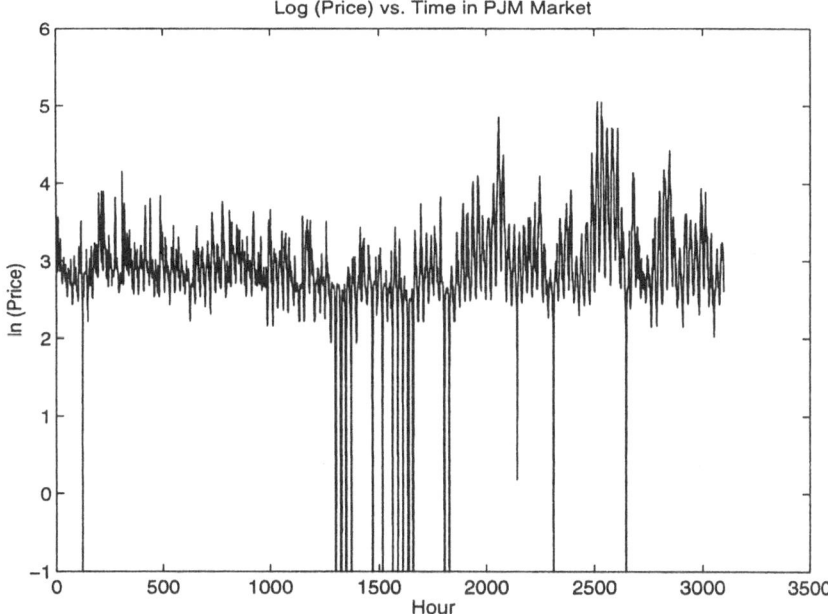

Figure 6.6: Plot of natural logarithm of price vs. time in the PJM electricity market. Hours for which the market clearing price is zero are plotted as having a log of −1.

Hours	Degrees of Freedom	η	$\overline{x}$	$\hat{\sigma}^2$
1 – 3096	3093	0.134	2.87	1.19×10^{-1}
125 – 1225	1098	0.171	2.90	2.58×10^{-2}
2649 – 3096	445	0.097	3.01	3.99×10^{-2}

Table 6.3: Parameter estimates for mean-reverting model of price logarithm.

Mean-reverting process

We now consider modeling the logarithm of price as a mean-reverting process. The regression results are given in Table 6.3. In order to allow for the possibility that the price process parameters may vary over time, the parameters will be estimated for each hour using weighted least squares, in which the squared residual for hour $i - 1$ has weight f with respect to the squared residual for hour i, where $0 < f \leq 1$. The regression can be updated for each hour using a recursive algorithm [31], as described in Appendix A. For the estimated variance, the appropriate bias factor is 1.020. Upon applying the mean reverting model for price prediction, the lowest prediction variance (31.78) occurs for $f = 0.99$; the standard deviation is therefore about 5.64.

Load regression line

The advantage of the first two price prediction methods is that they do not need any estimate of the load. However, they have the disadvantage of relatively large prediction error, which can be lowered if an accurate estimate of system demand is available. The remaining prediction methods assume that the load is known exactly; this information is then used to predict the price. A discussion of the effects of uncertainty in load prediction is deferred to the end of this chapter.

A simple method for using the system load to predict price is to simply apply the regression curve of price v. load. Using the estimate of variance from Table 6.4, the bias factor for the price estimate is 1.028. As before, the recursive regression algorithm with a "forgetting" factor may be used. For the PJM price data, setting f to 0.825 produces the lowest prediction variance; however, this variance value is 45.64, giving a standard deviation of 6.76. Although this method does allow for the use of load prediction, it does not accurately estimate the price.

Linear load function with intercept as random walk

We now consider the first of the two price process models that include load information. First, the parameters are estimated by linear regression, with the results shown in Table 6.5. The bias factor for this data is 1.015. Using

Hours	Degrees of Freedom	β_1	β_0	$\hat{\sigma}^2$
1 – 3096	3094	7.46×10^{-5}	0.755	2.74×10^{-1}
125 – 1225	1099	5.47×10^{-5}	1.525	4.26×10^{-2}
2649 – 3096	446	7.20×10^{-5}	0.741	5.52×10^{-2}

Table 6.4: Parameter estimates for linear load function model of price logarithm.

Hours	Degrees of Freedom	m	$\hat{\sigma}^2$
1 – 3096	3094	9.04×10^{-5}	1.07×10^{-1}
125 – 1225	1099	6.58×10^{-5}	1.99×10^{-2}
2649 – 3096	446	6.85×10^{-5}	3.00×10^{-2}

Table 6.5: Parameter estimates for linear load function model of price logarithm with intercept following random walk.

recursive least squares with a weighting factor $f = 0.999$, the variance of the error is 24.09, and the standard deviation is 4.91. Interestingly, this model performs better if the bias factor is neglected; with $f = 0.999$, the standard deviation of the prediction error is 4.82. The same observation was noted earlier in the pure random walk model. The bias factor does improve the prediction for the mean-reverting models, however.

Linear with function with mean-reverting intercept

The parameters for this model can not be estimated using linear regression, as equation (6.34) has four terms but only three independent variables. However, a numerical regression solution can be quickly calculated, as described in Appendix A. The resulting parameters are given for three time ranges in Table 6.6. Because a recursive algorithm is not available, the starting parameters were estimated from the first 11 hours of data, and the parameters were recomputed after every 100 hours, with the $(i - 1)$th squared residual weighted by f with respect to the ith residual. With $f = 0.99$ and a bias factor of 1.013, this method has a prediction variance of 21.15, or a standard deviation of 4.60.

Other possibilities

There are a number of other possible methods for modeling the price process of electricity. The method used in [30] is to find the mean for each day of the week during a given month, and then find a price process model for the price deviations from the mean. In [30], these deviations are modeled using

Hours	Degrees of Freedom	η	$\bar{b}$	m	$\hat{\sigma}^2$
1 – 3096	3092	0.215	0.504	8.34×10^{-5}	9.70×10^{-2}
125 – 1225	1097	0.262	1.345	6.18×10^{-5}	1.77×10^{-2}
2649 –.3096	444	0.317	0.788	7.05×10^{-5}	2.60×10^{-2}

Table 6.6: Parameter estimates for linear load function model of price logarithm with mean-reverting intercept.

an 8th-order autoregressive process. Another idea is to use a Fourier series to represent the mean as a periodic function of the time of year. However, as of this writing, there is only a few months of price data available from the newly formed PJM market, and this data is insufficient to provide parameter estimates for these methods. A very recent model in [5], postdating this research, uses a jump-diffusion process model, in which discrete price jumps occur according to a Poisson process.

6.3 Correlation of Load with Date

Since the price models with load information clearly improve the prediction of future prices, the value of good load estimates becomes clear. One factor which is correlated with the system load is the date of the year. Not surprisingly, the load on weekends is less than corresponding hours on weekdays, sometimes by as much as 20%. Using data from the PJM pool (Pennsylvania-Jersey-Maryland) for the years 1994 through 1996 [27], the ratio of the load on a given day of the week with respect to the average is shown in Table 6.7. In this table, each of the 24 hours in the day is tabulated separately. For example, on Monday morning from midnight to 1:00 AM, the average load is 95.8% of the average load for the hour from midnight to 1 AM over all days.

Another factor affecting load usage is economic growth. It should not be surprising to find a gradual increase in load from year to year, other factors being equal, because of new businesses and homes and population growth. In the PJM data, the average load growth from 1994 to 1995 was 1.97%, and the growth from 1995 to 1996 was 0.41%. The average annual growth rate over the entire period included in the PJM data is therefore about 1.19%. It should be noted that the higher growth rate for 1995 is partially influenced by the weather, as during the summer of 1995 the weather was quite hot, while during 1996 the summer was one of the coolest on record for the last 100 years. This observation also leads us to examine another potential determining factor for the amount of load: temperature.

Hour	Mon	Tue	Wed	Thu	Fri	Sat	Sun
12 AM - 1 AM	0.958	1.003	1.013	1.023	1.025	1.009	0.969
1 AM - 2 AM	0.966	1.004	1.014	1.024	1.025	1.003	0.965
2 AM - 3 AM	0.974	1.006	1.016	1.026	1.026	0.998	0.954
3 AM - 4 AM	0.980	1.008	1.017	1.027	1.026	0.991	0.952
4 AM - 5 AM	0.988	1.014	1.022	1.031	1.029	0.979	0.938
5 AM - 6 AM	1.004	1.031	1.037	1.045	1.041	0.945	0.897
6 AM - 7 AM	1.028	1.060	1.065	1.070	1.062	0.888	0.827
7 AM - 8 AM	1.039	1.069	1.074	1.078	1.072	0.871	0.797
8 AM - 9 AM	1.036	1.057	1.062	1.067	1.063	0.900	0.816
9 AM - 10 AM	1.031	1.046	1.052	1.058	1.053	0.923	0.836
10 AM - 11 AM	1.030	1.043	1.048	1.054	1.049	0.930	0.846
11 AM - 12 PM	1.029	1.042	1.048	1.052	1.046	0.928	0.855
12 PM - 1 PM	1.029	1.044	1.049	1.052	1.044	0.921	0.861
1 PM - 2 PM	1.031	1.049	1.054	1.055	1.045	0.909	0.856
2 PM - 3 PM	1.031	1.052	1.058	1.057	1.045	0.902	0.854
3 PM - 4 PM	1.032	1.053	1.060	1.057	1.041	0.899	0.858
4 PM - 5 PM	1.031	1.053	1.059	1.054	1.032	0.902	0.867
5 PM - 6 PM	1.032	1.050	1.057	1.050	1.022	0.907	0.881
6 PM - 7 PM	1.030	1.046	1.052	1.047	1.014	0.916	0.895
7 PM - 8 PM	1.030	1.043	1.050	1.046	1.006	0.919	0.906
8 PM - 9 PM	1.029	1.041	1.048	1.046	1.000	0.920	0.916
9 PM - 10 PM	1.023	1.035	1.043	1.042	1.000	0.928	0.928
10 PM - 11 PM	1.012	1.025	1.034	1.035	1.007	0.947	0.939
11 PM - 12 AM	1.003	1.017	1.025	1.028	1.010	0.964	0.952

Table 6.7: Ratio of load usage as a function of day of the week to average for a given hour in the day.

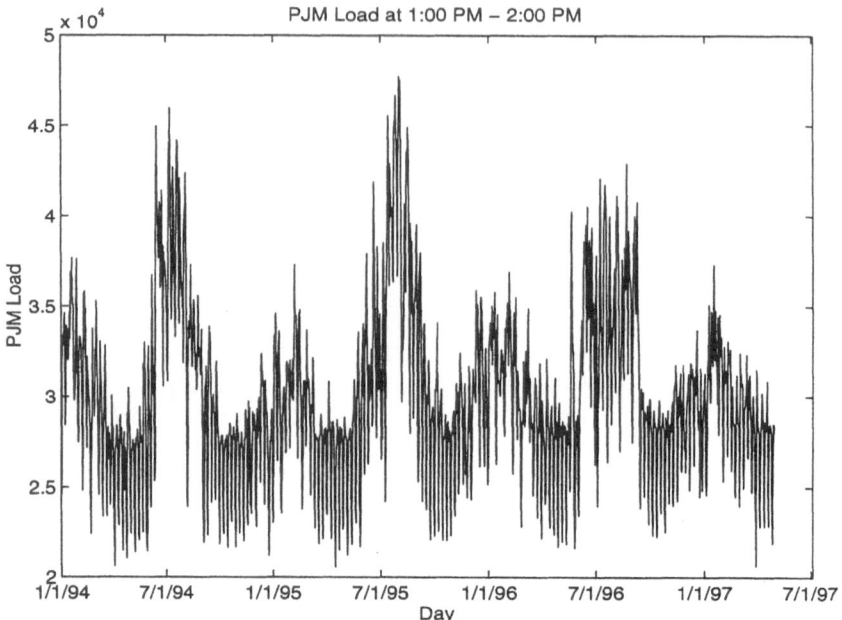

Figure 6.7: PJM load during the hour from 1:00 PM to 2:00 PM.

6.4   Correlation of Load with Temperature

It is generally well known that electricity usage is strongly influenced by the outdoor temperature. On hot days, heavy air conditioning usage results in a large load on the system, while electricity usage also is higher during cold weather for heating. A graph of afternoon load on the PJM pool (see Figure 6.7) reveals two peaks which occur in the summer and the winter, with the summer peak being noticeably higher. Furthermore, scatter plots of the PJM load with respect to the daily maximum at Philadelphia International Airport (PHL) [32] illustrate a strong correlation, although the relation is clearly nonlinear (Figure 6.8). (The temperature data covers the 912-day period from 1/1/94 to 6/30/96.)

In order to reduce the effects of day of week and annual growth on the load data when studying the relationship between load and temperature, an adjusted load $L_{adj}(t)$ is computed from the actual load usage $L(t)$ according to:

$$L_{adj}(t) = \frac{L(t)e^{-kt}}{R_w(t)} \tag{6.37}$$

where t is the time measured in hours, $k = 1.346 \times 10^{-6}$, representing a 1.19% annual growth rate, and $R_w(t)$ is the ratio of load to the average

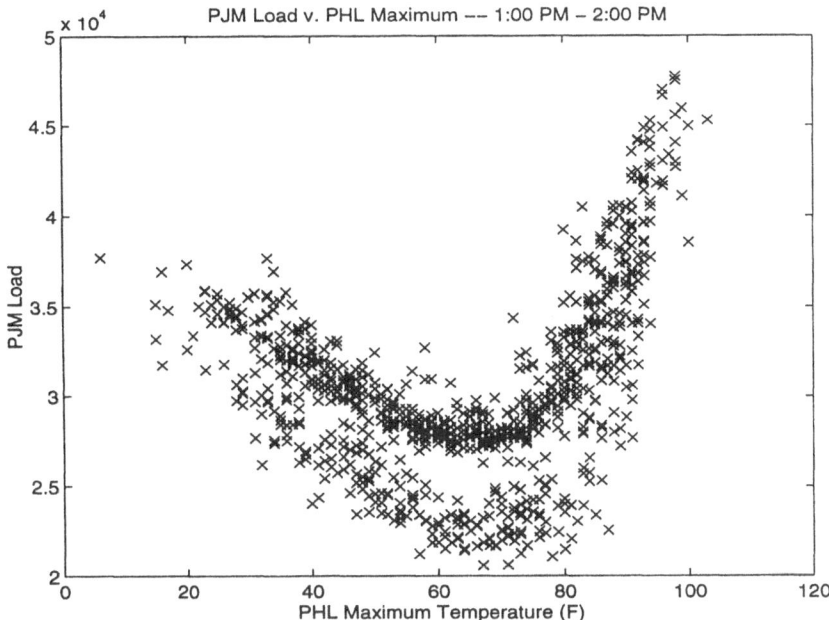

Figure 6.8: Scatter plot of load at 1:00 PM - 2:00 PM versus daily high temperature at Philadelphia International Airport.

for the hour of the day as a function of the day of week, as tabulated in Table 6.7. Additionally, the holidays of New Year's Day, Memorial Day, July 4th, Labor Day, Thanksgiving, and Christmas are estimated to have load use patterns similar to Sundays, and are adjusted accordingly in equation (6.37). The Friday after Thanksgiving is also observed to have reduced load and is adjusted using the Saturday weighting factor. The adjusted load is observed to provide an improved fit, as shown by the plot in Figure 6.9.

6.4.1   Overview of Linear Regression Analysis

Linear regression is the process of fitting a linear model to describe the relationship among two or more variables. The relationship is often used to predict future values of a quantity. Mathematically, the linear model for regression may be described by [29]:

$$Y = \beta_0 + \beta_1 X_1 + \beta_2 X_2 + \cdots + \beta_p X_p + e \qquad (6.38)$$

Here Y is known as the dependent variable or response, while the X_i are known as independent variables, and e is an error term. It is assumed that the error e has zero mean and a fixed variance σ^2, and the errors between different cases are uncorrelated. In regression, n observations of Y for a

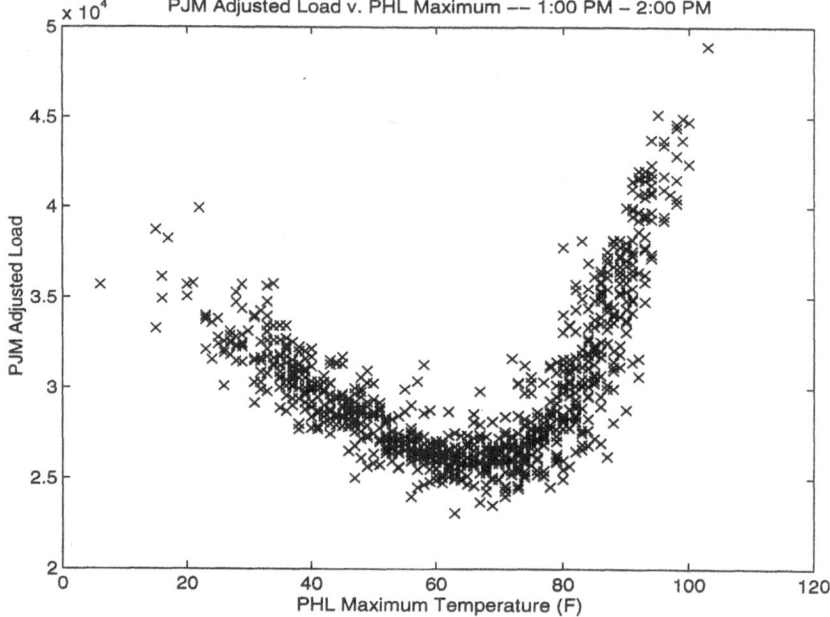

Figure 6.9: Scatter plot of adjusted load at 1:00 PM - 2:00 PM versus daily high temperature at Philadelphia International Airport.

corresponding set of X_i are given, and the object is to find the values for β_i that "best" describe the relationship between Y and X_i. The normal procedure is to select β_i to minimize the residual sum of squares, which will be defined shortly. Let y_k denote the value of Y and x_{ki} the value of X_i for the k-th observation. The complete set of data may then be written in matrix notation as [29]:

$$
y = \begin{bmatrix} y_1 \\ y_2 \\ \vdots \\ y_n \end{bmatrix}
\tag{6.39}
$$

$$
\beta = \begin{bmatrix} \beta_0 \\ \beta_1 \\ \vdots \\ \beta_p \end{bmatrix}
\tag{6.40}
$$

$$
X = \begin{bmatrix} 1 & x_{11} & x_{12} & \cdots & x_{1p} \\ 1 & x_{21} & x_{22} & \cdots & x_{2p} \\ \vdots & \vdots & \vdots & \ddots & \vdots \\ 1 & x_{n1} & x_{n2} & \cdots & x_{np} \end{bmatrix}
\tag{6.41}
$$

In this matrix notation, the rows correspond to individual cases or sets of observations, while the columns correspond to different independent variables. The regression model of equation (6.38) may thus be written as [29]:

$$
y = X\beta + e
\tag{6.42}
$$

where the elements of the error vector e are unobservable.

For an estimate $\hat{\beta}$ of β, the fitted values of y are denoted by $\hat{y}$ and defined as:

$$
\hat{y} = X\hat{\beta}
\tag{6.43}
$$

The residual vector, denoted here by r, is the difference between the observed and fitted values:

$$
r = y - \hat{y}
\tag{6.44}
$$

The residual sum of squares (RSS) is the sum of the squares of the elements of r:

$$
RSS = r^T r = \sum_{i=1}^{n} r_i^2
\tag{6.45}
$$

The least squares estimate $\hat{\beta}$ which minimizes RSS is [29]:

$$\hat{\beta} = (\mathbf{X}^T\mathbf{X})^{-1}\mathbf{X}^T\mathbf{y} \tag{6.46}$$

An estimate of the variance of e may be computed by dividing RSS by its degrees of freedom:

$$\hat{\sigma}^2 = \frac{RSS}{n-p-1} \tag{6.47}$$

The square root of this estimate $(\hat{\sigma})$ is known as the standard error of regression.

After a least squares computation is performed, there are a large number of tests and studies that can be done to determine whether the original linear model is in fact appropriate for the data. A fundamental test that is available tests whether the dependent variable Y is in fact related to any of the independent variables X_i. This test, known as the F-test for regression, compares the residual sum of squares with the total variability in Y, which is quantified by SYY, the total corrected sum of squares:

$$SYY = \sum_{i=1}^{n}(y_i - \bar{y})^2 \tag{6.48}$$

$$\bar{y} = \frac{1}{n}\sum_{i=1}^{n}y_i \tag{6.49}$$

The F-test is performed by computing the F-statistic:

$$F = \frac{SYY - RSS}{p\hat{\sigma}^2} \tag{6.50}$$

The test compares the model in equation (6.38) with the null hypothesis, which assumes that Y has no dependence on any of the X_i:

$$y_i = \beta_0 + e \tag{6.51}$$

If e is normally distributed with zero mean and variance σ^2, then under the null hypothesis of equation (6.51), F will follow an F-distribution with $(p, n-p-1)$ degrees of freedom. A large value of F which exceeds a chosen significance level strongly indicates that at least one of the coefficients of X_i in equation (6.38) is nonzero, since it is highly improbable that such a large value of F would be obtained under the null hypothesis model of equation (6.51). This result therefore means that Y is related to at least one of the independent variables [29].

Another useful regression statistic is the coefficient of determination, denoted as R^2. This statistic is given by the formula:

$$R^2 = \frac{SYY - RSS}{SYY} \tag{6.52}$$

Conceptually, equation (6.52) gives the proportion of the variability of Y which is explained by the regression on the independent variables. R^2 varies from 0 to 1, with larger values indicating a better regression fit [29].

The residuals can provide much information on the validity of the fitted model. By combining equations (6.44), (6.43), and (6.46), we find that the residuals are equal to:

$$\mathbf{r} = (\mathbf{I} - \mathbf{X}(\mathbf{X}^T\mathbf{X})^{-1}\mathbf{X}^T)\mathbf{y} \tag{6.53}$$

or, by defining the matrix V:

$$\mathbf{V} = \mathbf{X}(\mathbf{X}^T\mathbf{X})^{-1}\mathbf{X}^T \tag{6.54}$$

we have:

$$\mathbf{r} = (\mathbf{I} - \mathbf{V})\mathbf{y} \tag{6.55}$$

It can be shown from this last equation that all of the residuals have zero mean, and the variance of the i-th residual is $\sigma^2(1-v_{ii})$; however, the residuals are all correlated [29]. To provide a better basis for analysis, the residuals may be scaled by Studentization, or division by their estimated standard deviation. The i-th Studentized residual, denoted here as r_{si}, is given by:

$$r_{si} = \frac{r_i}{\hat{\sigma}\sqrt{1 - v_{ii}}} \tag{6.56}$$

where v_{ii} is the i-th diagonal element of $\mathbf{V}$. Although the Studentized residuals are not strictly t-distributed, they may be approximated as having a normal distribution with mean zero and variance one, if the linear regression model is true. Furthermore, the residuals are weakly correlated with the fitted values $\hat{y}_i$; this correlation is ignored for most analyses [29].

Once the Studentized residuals are calculated, they may be plotted as a function of the fitted values $\hat{y}_i$ to check the accuracy of the model. If the graph shows the residuals centered around zero with approximately 95% of the residuals between -2 and 2 for the entire range of $\hat{y}_i$, then there is no reason to suspect that the model is inappropriate. However, if systematic features are present, such as a change in the mean and/or variance of the residuals as a function of $\hat{y}_i$, then the model is likely inappropriate or incomplete for the data. Many large residuals may also indicate an inaccuracy in the model; with normally distributed errors, only about 3 out of 1000 residuals should lie outside the range -3 to 3. Residuals may also be plotted as a function of the independent variables X_i; these plots should also reveal no systematic features if the model is valid. A change in variance of the residuals indicates that the variance of the error is not constant but instead is a function of the other variables. If the mean of the residuals is not zero along the entire range of the plot, then the underlying process is likely nonlinear [29].

If the relation is suspected or known to be nonlinear (as in the load vs. temperature data shown previously), then polynomial regression may be used. Polynomial regression uses a polynomial of order d to represent the model [29]:

$$Y = \beta_0 + \beta_1 X + \beta_2 X^2 + \cdots + \beta_d X^d + e \qquad (6.57)$$

The coefficients $\beta_0, \beta_1 \ldots, \beta_d$ may be estimated through standard linear regression by treating $X, X^2, \ldots, X^d$ as separate independent variables [29].

6.4.2 Application of Regression to Temperature/Load Data

The relation between the load and temperature is estimated by regression of the load data with a cubic polynomial of the maximum temperature [29]. For the hour from 1:00 PM to 2:00 PM, the coefficient of determination of this regression is $R^2 = 0.6445$ using the unadjusted load data and $R^2 = 0.8334$ using the adjusted load data. Again, the adjusted data clearly provides a better fit to the temperature data. The F-statistic for the adjusted load data is 1514, while the 99% significance level for the F-distribution with $(3, 908)$ degrees of freedom (denoted $F(0.01; 3, 908)$) is approximately 3.8, providing overwhelming evidence that the load is a function of maximum temperature [29]. The 95% confidence interval for the coefficient of the cubic term is 0.128 to 0.153, strongly suggesting that this coefficient is non-zero. If a fourth power term in temperature is added to the regression, $R^2 = 0.8336$, which is almost identical to the cubic model, while the 95% confidence range for the highest order coefficient is from -2.91×10^{-4} to 6.88×10^{-4}, which does not exclude zero. These results therefore suggest that a third order polynomial in temperature is the best model to use for the regression.

The load/temperature relation may be further improved by adding a cubic polynomial of the minimum temperature for the day to the model. The resulting regression calculation for 1:00 PM to 2:00 PM gives $R^2 = 0.9074$, while the F-statistic is 1471, much greater than $F(0.01; 6, 904) \approx 2.8$. (Note that minimum temperature data for PHL is not available for four of the 912 days [32]; these days are excluded from the regression analysis.) The 95% confidence interval for the cubic coefficients of both maximum and minimum temperatures does not include zero, suggesting that this model provides a good basis for regression.

6.4.3 Conclusions of Regression Analysis for Load vs. Temperature

The complete regression results for each of the 24 hours in the day are provided in Appendix B. In the morning hours from 12:00 AM to 10:00 AM, it is observed that the previous day's high temperature provides a better fit in the regression than does the current day's maximum. A study of the residual plots and outlying cases leads to several conclusions. First, the cubic model may not be appropriate for extremely high or low temperatures, as systematic errors are observed in many residual plots for very high values of expected adjusted load. Specifically, the cubic model appears to overestimate the load if the temperature is very high; the load does not appear to grow as a cubic function of temperature. Secondly, although the variance appears to be relatively constant in the residual plots, an unusually large number of cases with high residuals (outside ±3) are observed. The large errors for these cases appear to be caused by unusual weather conditions: normally, the maximum temperature occurs in early to mid afternoon while the minimum temperature occurs in the early morning just prior to sunrise. On some days, however, this is not the case, and for these days, the daily maximum and minimum suggest that the temperature (and load usage) at the hour of study is different from the actual temperature. The availability and use of hourly temperature data would likely eliminate this problem. Other possibilities are the loss of power for an unusually large number of customers, possibly due to the presence of thunderstorms, and a few unusual days where temperatures across the PJM service area are highly variable.

6.5 Effects of Variance of Load Estimation

Since the load can never be predicted exactly, it is important to determine whether the error in load estimation outweighs the improved price forecasts from models that use a demand estimate. In order to answer this question, we begin by assuming that price deviations and errors in load estimation are uncorrelated. While this may not be strictly true (particularly for catastrophic events), it seems reasonable to believe that these two processes generally act independently of each other. Given this assumption, the additional variance in the logarithm of price (denoted σ_{LE}^2) may be added to the variance of the price process models to obtain the total variance. The variance of $\ln P$ due to load estimation in all of the models using load prediction is:

$$\sigma_{LE}^2 = \text{var}(mL) = m^2 \text{var } L \tag{6.58}$$

The value of σ_{LE}^2 should not exceed a threshold. This threshold is the difference in variance between the "best" model that includes load prediction and the "best" model that does not. For the PJM data, this threshold is hard to define, as the variances differ significantly over different price ranges. Also, the value of m^2 varies over a large range. If the threshold is set at 0.01 and m is approximately 1×10^{-4}, the standard error of the load estimate should not exceed 1000. For $m = 8 \times 10^{-5}$, the standard error may be as much as 1250. If the threshold is increased to 0.022 (the variance difference when all price data is included), the standard error for the load estimate can reach 1854 if $m = 8 \times 10^{-5}$.

Using the temperature data for price prediction appears to have marginal value, at least for the regressions performed in this book. The standard errors for the load are in the same range as the maximum limits estimated in the previous paragraph (without considering the variance of temperature forecasts). It is interesting to note that the estimates for m are lowest for time ranges in which the variance (and variance threshold) is lowest. The standard error of the load estimate may also be improved with additional information. Clearly, however, the price models that include a load estimate will provide useful information about price if tomorrow's forecast calls for a sudden jump in temperature from 60° F to 95° F; price models without a load forecast will not capture the expected price increase.

Although it is not clear that the mean-reverting intercept price model with load estimates offers significantly better prediction than the standard AR(1) model, this model is nevertheless used as the price model for the remainder of this book. The mean-reverting intercept model can be used to capture the changes in average price over different hours of the day, as depicted by Figure 6.1. This price model also makes it possible to determine optimal unit commitment decisions as a function of time of day, on the important assumption that the price is at its average value.

CHAPTER 7
COMPUTATIONAL COMPLEXITY OF UNIT COMMITMENT

In previous chapters, a model for the evolution of the price of power was developed, taking into account price and demand forecasting techniques. This chapter addresses the question of how to use the available information in order to make an optimal commitment decision at each hour.

The addition of the price process model adds a state variable to the unit commitment problem. In addition to x_k, b_{k-1}, which is the intercept quantity in the price model from the previous hour, is also a state variable. u_k remains the only control variable, while π_k, the profit during hour k, is a random quantity given by equations (3.2) and (3.3), which are restated here for convenience:

$$P_G = \frac{p_k - b}{2a} \tag{7.1}$$

$$
\begin{aligned}
\pi_k \ = \ & u_k(p_k P_G - c_G(P_G) - I(x_k < 0)S) \\
& - (1 - u_k)(c_f + I(x_k > 0)T)
\end{aligned} \tag{7.2}
$$

The price p_k is a function of the load L_k and the price state b_k, from equation (6.30):

$$\ln p_k = mL_k + b_k \tag{7.3}$$

while the state b_k evolves according to equation (6.34):

$$b_k = \bar{b}(1 - e^{-\eta}) + e^{-\eta}b_{k-1} + e_k \tag{7.4}$$

The state transition equation for x_k is [3]:

$$x_{k+1}(1) = \begin{cases} \max(1, x_k + 1) & : u_k = 1 \\ \min(-1, x_k - 1) & : u_k = 0 \end{cases} \tag{7.5}$$

There are two random inputs to the system; the error term e_k in the price model, and the error of the estimation of the load L_k.

7.1 Dynamic Programming Algorithm

It is generally well recognized that dynamic programming is applicable to many stochastic optimization problems; however, it also has the disadvantage of non-polynomial (NP) growth of operation count with respect to problem size (the "curse of dimensionality") [1]. The unit commitment algorithm developed in [2] for a coordinated utility has a deterministic problem formulation, as the number of control choices is equal to 2^{N_G}, where N_G is the number of generators. However, the problem for an owner of a single generator has only two control choices (on or off); therefore, it becomes feasible to address unit commitment in a stochastic environment.

A direct implementation of dynamic programming for the unit commitment problem with a price process and generation limits may be used. Equation (4.4) for the unit commitment problem becomes:

$$J_N(x_N, b_{N-1}) = 0 \tag{7.6}$$

$$J_k(x_k, b_{k-1}) = \max_{u_k \in \{0,1\}} \underset{p_k}{E} \{\pi_k(x_k, u_k, p_k) + J_{k+1}(x_{k+1}, b_k)\} \tag{7.7}$$

The computation of the optimal decision is as follows: Starting at the last time stage, the optimal profit and optimal decision are calculated for every possible state. Next, the optimal profit-to-go and decision are calculated for every state in the next to last stage, using the optimal profit information from the last stage. This process continues until the first (current) time stage is reached.

Because the state includes the price intercept b_{k-1}, which is a continuous variable, the states must be discretized in order to carry out the above procedure. The difference between two consecutive discretized states will be denoted as d. The range of price intercepts considered at the first stage will encompass 5 standard deviations above and below the initial value b_{-1}. Intercept values outside this range at the first time stage are highly improbable and are assumed to have negligible impact on the expected profit. At the second time stage, the intercept range considered is 10 standard deviations above and below b_{-1}, so that the expected profit-to-go at the first stage can be calculated using a range of 5 standard deviations. The n-th stage has a range of $5n$ standard deviations above and below the starting intercept of b_{-1}.

The first time stage has $2(5\sigma d^{-1}) + 1$ price states. Each succeeding stage adds $2(5\sigma d^{-1})$ price states. Therefore, the N-th stage (where N is the horizon length) has $2N(5\sigma d^{-1}) + 1$ price states. However, states relating to the generator status (on or off) must also be included. The number of these states for one generator is $(t_{up} + t_{dn})$, the sum of minimum up and down times. The total number of states is the product of these quantities. The n-th stage has:

$$(t_{up} + t_{dn})^{N_G}(10n\sigma d^{-1} + 1)$$

total states, where N_G is the number of generators. Recall that an optimal profit-to-go and optimal decision must be calculated for each state. This computation requires calculating the expected profit for one stage and then adding the expected profit-to-go from the next stage, which is approximated by multiplying the transition probability to a given state by the expected profit-to-go for that state, and then summing over all possible future states. If more than one control choice is allowed, this computation is performed for each control choice, and the highest resulting profit determines the optimal control. Note that the total number of states to consider in a time horizon of N stages is:

$$(1/2)N(N + 1)(t_{up} + t_{dn})^{N_G}(10\sigma d^{-1}) + N(t_{up} + t_{dn})^{N_G}$$

7.2 Heuristic Simplifications

In many cases, the exhaustive computation of an optimal unit commitment decision gives a result which is intuitively obvious. For example, at 3 PM demand is typically high and remains high for several hours; since the price would be expected to be correspondingly high, it would be expected that most typical low or moderate cost generators should be turned on (if they are not already on). Such intuitive reasoning can be used to develop heuristic algorithms that cover many situations, although extensive computation will be necessary to cover most borderline cases. The primary difficulty with heuristic algorithms is that intuition is occasionally deceptive, and consequently many heuristic methods are not optimal, while it is very difficult to prove the optimality of other heuristic algorithms, even in specialized cases.

A good illustration of this difficulty in the decentralized unit commitment algorithm is the following heuristic: If the generator is running and the next hour is expected to be profitable, stay running. This algorithm will work for most practical situations; as shown earlier in Chapter 6, the price usually stays low throughout the night hours and rises during the day. For the example shown at the end of this chapter, the generator is generally turned off from 11 PM to 7 AM and is on at all other times. In this case, when the generator is turned off, it is expected to remain off for many hours. However, there is a counterexample to this algorithm: if a small profit is expected in the next hour, a large loss is expected in the following hour, and an even bigger profit is expected in the third hour, the heuristic described above will not be optimal for a minimum down time of 2 hours. This heuristic would require the generator to stay on for all three hours and absorb the large loss during the second hour. Instead, it would be optimal to turn off and forego the small profit of the first hour in order to avoid the large loss of the second hour. The generator can be restarted in order to capture the very large profit during the third hour.

7.3   Ordinal Optimization

Ordinal optimization [13, 14] is an approach for finding a solution to an optimization problem in which a large number of possible policies have to be considered. In ordinal optimization, the object is not to find the one best policy choice but rather to select a policy which, with very high probability, is among the highest percentile of all possible policies. This characteristic of ordinal optimization is termed goal softening; although it is obviously more desirable to find the very best solution, it may not be worthwhile or feasible to examine every possible solution out of a billion (or more) possibilities. Ordinal optimization is a prime candidate for problems in which the number of possible policies increases exponentially or combinatorially (such as the traveling salesman problem), the policy space has little or no structure, and evaluation of the objective criterion for each policy is corrupted by large noise or otherwise computationally expensive. Stochastic optimization problems have these characteristics, and therefore they are well suited to ordinal optimization.

7.3.1   Overview of Ordinal Optimization

The idea of goal softening is much more appropriate for some problems than for others. With high probability, ordinal optimization finds a "good enough" solution, defined as being among the top $n\%$ of all solutions. This does not, however, mean that a "good enough" solution is within $n\%$ of the optimal cost. For many problems, a fairly large number of solutions perform close to the true optimum. In these cases, the qualitative difference among "good enough" policies is small and more than offset by the expense of trying to find the one best solution. Other problems, however, may have one solution that stands out as being far ahead of the others; for such problems (termed "needle in a haystack" problems), goal softening is of little value. However, ordinal optimization can be used for these problems to learn about the characteristics of the better solutions and thus decide where to search next.

Ordinal optimization has three basic steps. First, a uniform sample of N policies out of all possible policies is taken. Second, using a selection rule, a subset of the N policy samples is picked. The size of this subset is sufficiently large such that at least one "good enough" solution is included in the selected subset with high probability. Finally, the policy choices in the selected subset can be evaluated more thoroughly, and the best policy of these is chosen.

The first step of ordinal optimization is to select a sample of the policy space. Ordinal optimization assumes that it is possible to sample the policy space uniformly; each policy has equal probability of being selected. The sample should be large enough so that we are virtually assured of having at least one of the top $n\%$ of the overall search space in the sample of N policies. This probability is calculated by the formula:

$$P = 1 - (1 - n\%)^N \qquad (7.8)$$

In [14], a sample size of $N = 1000$ is used; the resulting probability P that one of the top 5% is included is $1 - 5.29 \times 10^{-23}$, a virtual certainty. For $n\% = 1\%$, P is equal to 0.999957, which is not a guarantee, but is highly probable.

The next step is to choose a subset from the sample of N policies. This subset is denoted by S and contains s elements. A "good enough" subset G of the sample can also be defined as the top $n\%$ of the sample; G therefore has $g = N \times n\%$ elements. The question is how to choose the subset S and how many elements to include. This question is answered by determining the alignment probability P_A, which is the probability that the sets G and S have at least k elements in common. k is termed the alignment level. s is chosen so that the alignment probability exceeds a desired threshold for a selected percentile n (typically $n\% = 5\%$).

The selection rule is the process by which s elements are selected from N samples. Although there are many possible selection rules which can be used, two of the simplest methods are blind pick (BP) and horse race (HR). Blind pick is equivalent to sampling without replacement; s out of N policies are chosen, where each element has equal probability of being chosen. Note that under blind pick, no evaluation of any of the N policy choices is done. The alignment probability under blind pick can be written mathematically as:

$$P_A(k, s, g | BP) = \sum_{i=k}^{\min(g,s)} \frac{\binom{g}{i}\binom{N-g}{s-i}}{\binom{N}{s}} \qquad (7.9)$$

This value of alignment probability is known as the blind pick lower bound (BPLB); it is a lower bound since the selection method uses no knowledge of the cost of any of the policies. If some estimate of cost is available and this knowledge is used when selecting elements, then the alignment probability will be higher.

An obvious choice of selection method using cost estimates is known as horse race (HR). Under this method, all N policies are evaluated by finding an estimate of the cost for each policy. The actual method of doing the evaluations is problem dependent, although for most stochastic problems a Monte Carlo simulation would typically be used. The use of a relatively crude model for the horse race simulations leads to a significant reduction in computation. After completing the simulations, the s policies with the lowest cost are selected. Since each policy's simulated cost is an estimate of the true expected cost plus some noise ω_i:

$$\tilde{J}_i = J_i + \omega_i \tag{7.10}$$

the horse race results should improve the alignment probability with respect to the BPLB above. The amount of improvement depends on the noise variance and the shape of the ordered performance curve (OPC), which is defined in the next paragraph. Note that the ordering in the horse race simulation will not in general be identical to the actual ordering of the policies, since one policy may have a large positive noise element while another policy which is on average worse may appear to be better because of large negative noise.

The ordered performance curve is obtained by plotting the expected cost of each policy, in order from the lowest to the highest. The plot of discrete points may be approximated as a nondecreasing curve, with the x axis indicating the rank of a policy and the y axis indicating expected cost. Furthermore, the OPC may be standardized by scaling the $x-$ and $y-$axes to a range of $[0, 1]$. In the standardized OPC, the best policy is denoted by 0 and has an expected cost of 0 while the worst policy is denoted by 1 and has an expected cost of 1. All of the intermediate policies are represented by real numbers between 0 and 1, with higher ranking policies having lower numbers.

The reason for scaling the OPC curve is to classify its shape into one of five classes. In [14], the five classes are represented by using a smooth curve with only two parameters, α and β. Generally speaking, α is the slope of the OPC for the lower cost ("good") designs, while β is the slope of the OPC for the higher cost ("bad") designs. Table 7.1 lists the five classes and the corresponding range of α and β; the OPC shapes for each class are shown in Figure 7.1. For a given problem, the OPC curve is not known exactly (otherwise the problem is solved!); however, its class may be guessed by using intuition or a rough estimation; the neutral class is a good choice if no other evidence is available.

The OPC classes are used to estimate the alignment probabilities using the HR (horse race) selection method. A closed form expression for the alignment probabilities under HR is not available; however, an estimate of these probabilities from Monte Carlo simulation is given in [14], in which the

No.	Class	Category	Parameter Values
1	Flat	Lots of good designs	$\alpha < 1, \beta > 1$
2	U-Shaped	Lots of good and bad designs, but few intermediate designs	$\alpha < 1, \beta < 1$
3	Neutral	Good, intermediate, and bad designs are equally distributed	$\alpha \approx 1, \beta \approx 1$
4	Bell	Lots of intermediate designs, but few good and bad designs	$\alpha > 1, \beta > 1$
5	Steep	Lots of bad designs	$\alpha > 1, \beta < 1$

Table 7.1: The five classes of the ordered performance curve density.

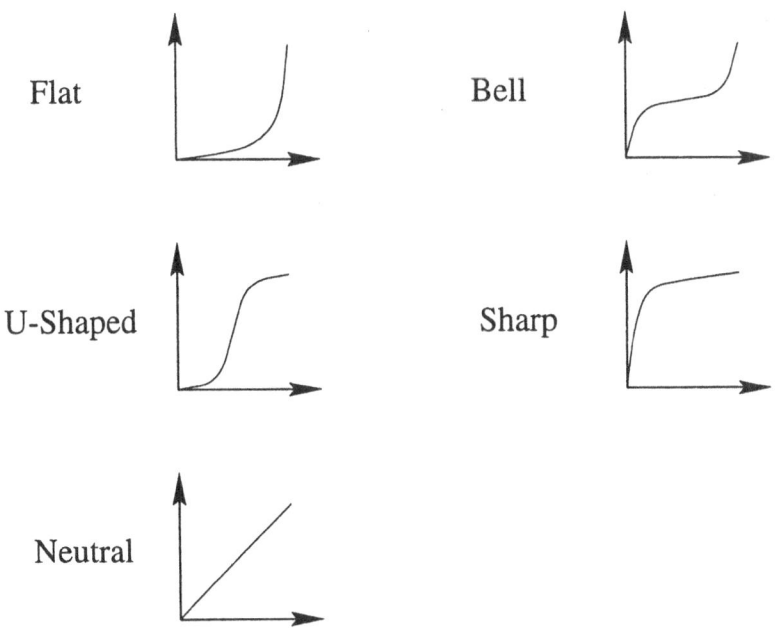

Figure 7.1: The five classes of ordered performance curve shapes.

estimates are used to determine $Z(k,g)$, which is defined as the minimum size of the selected subset required to achieve an alignment probability $P_A = 0.95$ for a given alignment level k and "good enough" criterion g. The Monte Carlo estimates of Z are fitted using regression to the following function, which is useful for all five OPC classes:

$$Z(k,g) = e^{Z_0} k^\rho g^\gamma + \eta \tag{7.11}$$

The parameters Z_0, ρ, γ, and η depend on the OPC class and also the variance of the noise ω_i in the horse race estimates of cost. Assuming that $20 \le g \le 200$, $Z < 180$, and k/g is small, equation (7.11) will accurately estimate how many elements should be selected from the sample of N policies in order to obtain k "good enough" solutions with 95% probability. Note that if the OPC is perfectly flat (meaning all policies are equally good) or the noise variance is infinite, then the horse race method offers no improvement over blind pick, since the policy rankings from a horse race outcome are uniformly random.

In summary, ordinal optimization may be performed for a given problem using a basic three-step procedure. First, a sample of N policies out of all possible choices is selected, using uniform random sampling. [14] recommends $N = 1000$ as being a sufficiently large sample; higher values of N improve the chance of obtaining exceptionally good policies, but the heavy computation required using larger samples is contrary to the spirit of ordinal optimization. Second, a subset of s elements is drawn from the sample of N policies. For most problems, the elements chosen are the s policies that have the lowest cost after all N sample policies are simulated using a Monte Carlo approach. The value of s is derived from equation (7.11) using an estimate of the OPC class of the problem (or the neutral class, if no better choice can be determined), and an estimate of the amount of noise present in each Monte Carlo simulation. The s selected policies may then be evaluated more thoroughly to determine which one offers the best performance. Ordinal optimization is a method for greatly reducing the search space for problems in which a thorough evaluation of all possible choices is impractical.

7.3.2 Application to Unit Commitment

In order to reduce the computation of the optimization algorithm, an ordinal optimization approach may be used. This approach is aided by using the conjecture that the optimal policy is a threshold policy with respect to the current price; if it exceeds a fixed threshold value, then it is optimal to turn on (or stay on); otherwise, the generator should be off. (Note, however, that the threshold depends on the current state and will in general vary from one hour to another.) While this conjecture is intuitive and supported by the numerical examples solved in this book, a proof is left for further research.

Ordinal optimization can be implemented by noting that any policy may be specified by providing two price thresholds for each hour in the planning horizon. One threshold is for the "on" state, while the other is for the "off" state; note that a decision is possible only if the state x_k is equal to t_{up} or $-t_{dn}$. The threshold for the "off" state will always exceed or equal the threshold for the "on" state because of the startup cost and minimum up time. For a price process model with load information, the threshold may also be expressed in terms of the intercept value; in this case, the actual threshold value for the price at a given hour depends on the load for the preceding hour, whereas a threshold value for the intercept may be calculated many hours in advance. A random policy can be determined by drawing $2N$ random numbers. For each pair of numbers, the lower number is the "on" state threshold for a given stage while the higher number will be the "off" state threshold for the same stage. After drawing a sample of 1000 random policies, each policy is evaluated by a Monte Carlo simulation, having knowledge of the initial state and price. An approximation of the ordered performance curve for the unit commitment problem is shown in Figure 7.2; this curve was derived by randomly selecting 1000 policies and running 500 Monte Carlo simulations per policy. Note that the OPC for unit commitment clearly falls into the "bell" class.

Alternatively, because of the principle of optimality [1], the problem may be tackled by backwards recursion. Beginning at the last stage, a subset of policies may be randomly selected and simulated. The best performing policy is then assumed to be the optimal policy, and the process moves back one stage, by selecting a new subset of thresholds for the next to last stage. Each simulation at a given stage is performed by using the assumed best policy for all future stages. If the number of policies at each stage is $1000/N$, then a total of 1000 simulations is performed, but the results may be expected to be more accurate, since in effect $(1000/N)^N$ policies are evaluated. The same random price path for a Monte Carlo simulation at a given stage may be used simultaneously for evaluating all policies; although no theoretical alignment probabilities are available, the resulting correlation in policy costs may be expected to keep the policies closer to their actual order [14].

7.4   Example

To demonstrate how the unit commitment algorithm under deregulation may be implemented, a hypothetical example is developed in this section. The cost of generation is equal to $c_G(P_G) = 2P_G^2 + 2P_G + 18$. While the generator is down, the fixed cost per hour is $c_f = 4$. The minimum up time is 3 hours and the minimum down time is 2 hours. The startup cost S is 4 and the shutdown cost T is also 4. The problem will be solved for a time horizon of $N = 24$ hours. The generation limits are $5 \leq P_G \leq 8$. The expected load for each of the next 24 hours is shown in Table 7.2. The price process is modeled using the mean-reverting intercept model from Chapter 6, with parameters

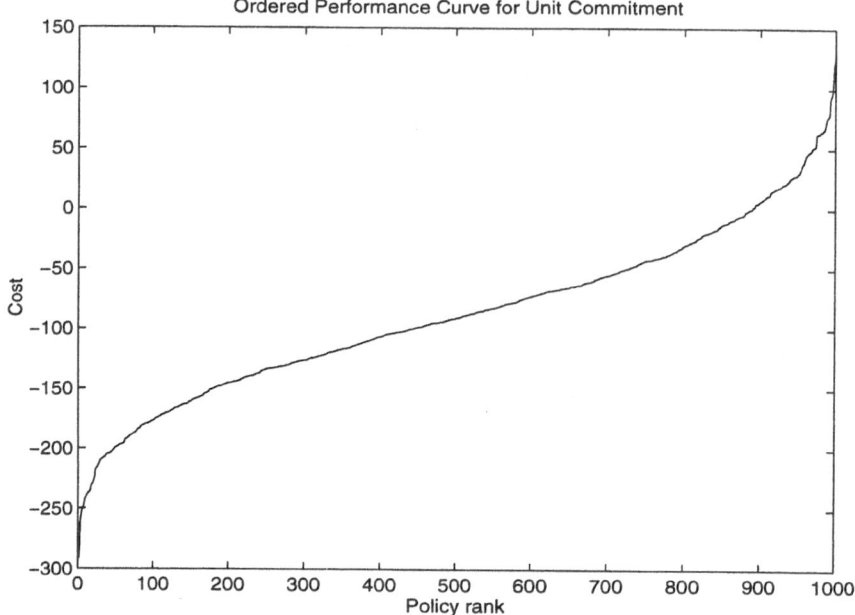

Figure 7.2: Simulated ordered performance curve of unit commitment policies.

Hour	Expected Load	Standard Deviation
12:00 AM	23830	996
1:00 AM	22402	955
2:00 AM	21531	925
3:00 AM	21291	909
4:00 AM	21374	903
5:00 AM	22429	948
6:00 AM	24562	1111
7:00 AM	27274	1177
8:00 AM	29123	1137
9:00 AM	30396	1149
10:00 AM	31462	1198
11:00 AM	31989	1242
12:00 PM	32089	1277
1:00 PM	32252	1332
2:00 PM	32275	1386
3:00 PM	32145	1438
4:00 PM	32138	1547
5:00 PM	32024	1829
6:00 PM	31448	1748
7:00 PM	30982	1570
8:00 PM	31216	1292
9:00 PM	31092	1221
10:00 PM	28937	1185
11:00 PM	26167	1134

Table 7.2: Load data for unit commitment example.

$\eta = 0.317$, $\bar{b} = 0.788$, $m = 7.05 \times 10^{-5}$, and $\sigma = 0.1612$. The starting price (from the previous hour) is $p_{-1} = 13.91$ with a load of $L_{-1} = 26167$.

7.4.1   Solution by Dynamic Programming

Using the exhaustive dynamic programming algorithm, with a discretization of 0.05 for the logarithm of price, the optimal decisions for 10:00 PM and 11:00 PM for the given data are shown in Tables 7.3 and 7.4. The optimal decision at 10:00 PM is to keep the generator in its current state. The optimal decision at 11:00 PM is to always turn off, if the minimum up time constraint is not violated. The DP algorithm required 84.307s of user time (101.95s of elapsed time) to calculate the solution for 10:00 PM. If the problem is solved for each hour of the day using the load data from Table 7.2, with the previous hour's intercept at the mean value $\bar{b}$, the resulting unit commitment

Status	Expected Profit	Optimal Decision
On since 9:00 PM	397.21	On
On since 8:00 PM	402.42	On
On since 7:00 PM	402.42	On
Off since 9:00 PM	391.28	Off
Off since 8:00 PM	391.28	Off

Table 7.3: Optimal decision and expected profit for each state at 10:00 PM

Status	Expected Profit	Optimal Decision
On since 10:00 PM	373.66	On
On since 9:00 PM	387.38	On
On since 8:00 PM	390.36	Off
Off since 10:00 PM	394.37	Off
Off since 9:00 PM	394.37	Off

Table 7.4: Optimal decision and expected profit for each state at 11:00 PM

decisions are shown in Figure 7.3. Note that a unit commitment schedule under this problem formulation is not drawn up for a 24-hour period; instead, a unit commitment decision is made at each hour. Price jumps or drops that occur in the near future will affect the optimal decision when those hours are reached.

7.4.2 Solution by Ordinal Optimization

Several methods of applying ordinal optimization to the unit commitment problem were outlined earlier in Section 7.3.2. The most direct method is to randomly select 1000 policies and simulate them. The thresholds of the intercepts are drawn uniformly from $[0, 2]$; a suitable range may be found by repeated application of ordinal optimization. Each policy is simulated 25 times with both possible initial decisions (on and off). The top performing policies, starting at 10:00 PM, are shown in Table 7.5. The average simulated profit and its standard deviation is shown for each policy. The highest observed standard deviation in the simulations is 80. Since the cost range in the OPC (Figure 7.2) between the best and worst solutions is at least 450, the medium noise range class may be used to estimate $Z(k, g)$. From [14], the parameters in equation (7.11) are $Z_0 = 8.1998$, $\rho = 1.9164$, $\gamma = -2.0250$, and $\eta = 10.00$; to obtain one of the top 5% policies with 95% probability, 12 policies are sufficient ($Z(1, 50) = 11.3$).

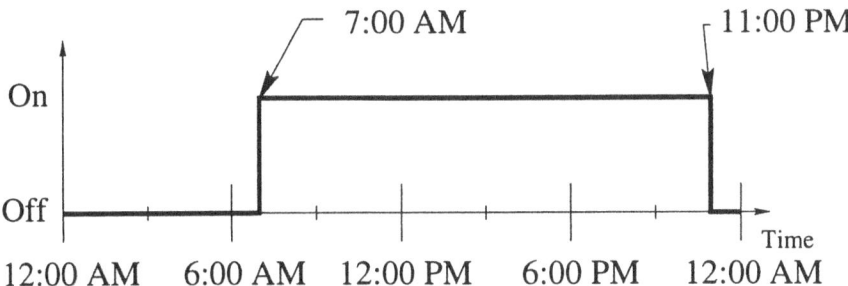

Figure 7.3: Optimal unit commitment decision over a 24 hour period.

	Off		On		Optimal
No.	Profit	Std. Dev.	Profit	Std. Dev.	Decision
1	351.07	60.93	230.77	41.42	Off
2	240.57	38.97	349.45	64.23	On
3	341.43	49.37	266.55	59.43	Off
4	233.35	62.62	324.45	47.37	On
5	309.40	72.37	235.94	47.73	Off
6	130.26	34.64	302.91	66.41	On
7	217.23	52.63	301.49	57.18	On
8	236.63	52.19	294.23	31.64	On
9	293.83	40.14	206.54	33.40	Off
10	290.00	80.10	123.09	37.07	Off
11	284.03	47.01	222.42	45.35	Off
12	283.12	51.30	182.03	51.87	Off

Table 7.5: Results of ordinal optimization of 1000 random policies.

Without significantly increasing the computation, it appears that this implementation of ordinal optimization is not very satisfactory. There is significant disagreement among the selected policies regarding the optimal decision. The simulation noise appears to be a greater influence on the optimal decision that the choice of policy. Another problem with this algorithm is that any selected policy still generally has at least one time stage where the threshold is far from optimal. Comparison of the thresholds of the policies selected by the ordinal optimization algorithm (Tables 7.6 through 7.8) with the optimal policy in Table 7.9, extracted from the dynamic programming algorithm, illustrates this fact. Using identical price samples for both the on and off simulations (as mentioned in Section 7.3.2) could alleviate these problems, although the backwards iteration algorithm would appear to make much better use of the same number of computations.

Hour	Policy # 1 On	Policy # 1 Off	Policy # 2 On	Policy # 2 Off	Policy # 3 On	Policy # 3 Off	Policy # 4 On	Policy # 4 Off
1	0.08	1.83	0.24	0.86	0.07	1.57	0.61	1.46
2	0.60	0.68	0.25	0.35	0.10	1.33	0.28	0.29
3	1.17	1.65	0.17	0.29	1.32	1.55	1.17	1.19
4	1.39	1.87	0.30	0.41	0.53	1.40	1.14	1.56
5	0.02	0.45	0.79	1.73	0.83	1.34	1.24	1.41
6	1.47	1.81	0.39	0.52	0.66	1.75	0.08	0.82
7	0.02	1.48	1.32	1.74	0.66	0.91	0.94	1.49
8	0.66	0.74	0.37	1.57	0.18	0.52	0.27	1.78
9	0.37	1.44	0.22	0.27	1.55	1.67	0.02	0.49
10	0.29	0.88	0.87	1.88	0.02	1.84	0.72	0.88
11	0.35	1.84	0.70	1.91	0.23	0.84	0.85	1.19
12	0.41	1.10	0.49	1.57	0.15	0.41	0.14	0.62
13	1.42	1.45	1.13	1.97	0.60	1.76	0.38	1.89
14	1.25	1.27	0.75	1.90	0.16	0.37	0.17	0.96
15	0.69	0.79	0.20	0.46	0.62	1.23	0.08	1.02
16	0.36	0.53	1.09	1.30	0.20	0.30	0.18	0.38
17	0.21	1.17	1.51	1.79	0.31	0.68	0.04	1.37
18	0.17	0.44	0.47	0.97	0.84	1.62	0.21	1.32
19	0.07	0.47	0.14	0.21	1.52	1.54	0.61	1.07
20	0.05	0.37	0.38	0.97	0.56	1.11	0.89	1.22
21	0.19	1.86	0.06	0.99	0.24	0.33	1.56	1.91
22	0.30	1.49	0.46	1.08	1.09	1.21	1.19	1.51
23	0.10	0.44	0.85	1.06	0.83	1.03	0.67	1.82
24	0.27	0.81	1.18	1.87	1.95	1.99	1.21	1.34

Table 7.6: Thresholds for policies #1 - #4.

Hour	Policy # 5 On	Policy # 5 Off	Policy # 6 On	Policy # 6 Off	Policy # 7 On	Policy # 7 Off	Policy # 8 On	Policy # 8 Off
1	1.69	1.95	0.33	0.38	0.95	1.22	1.41	1.50
2	1.55	1.90	0.59	1.09	0.46	1.98	1.69	1.79
3	0.59	1.80	0.78	1.19	0.01	0.87	0.42	1.94
4	0.76	1.26	0.77	1.45	0.61	1.59	0.70	1.15
5	1.17	1.92	1.56	1.91	0.22	1.63	0.34	1.80
6	1.59	1.78	0.58	1.98	0.31	0.90	0.49	0.94
7	0.02	0.06	0.89	1.21	0.76	0.94	1.02	1.52
8	1.56	1.85	0.42	0.85	0.24	1.29	0.54	1.13
9	1.41	1.79	0.30	0.76	0.61	1.40	0.41	1.96
10	0.04	0.88	0.01	1.21	0.56	1.77	1.13	1.59
11	0.51	1.14	1.01	1.35	0.29	0.71	0.94	1.98
12	0.44	1.99	0.28	0.92	1.08	1.18	0.53	0.94
13	0.63	1.60	0.08	0.20	1.31	1.84	0.07	0.21
14	1.18	1.32	0.37	1.35	0.46	0.50	0.28	0.55
15	0.82	1.71	0.68	1.49	0.28	0.94	0.99	1.84
16	0.49	0.60	1.02	1.19	0.21	0.67	0.18	0.21
17	0.23	1.73	0.50	1.12	0.62	0.98	0.38	1.81
18	0.23	1.50	0.83	0.84	0.73	0.89	0.59	0.97
19	0.35	0.56	0.82	1.53	0.64	0.73	0.61	1.99
20	0.77	1.17	0.14	1.79	0.40	1.54	0.08	1.16
21	0.24	0.64	0.06	1.36	0.60	1.15	0.52	0.70
22	1.85	1.88	0.60	1.02	0.53	0.83	0.23	1.12
23	0.84	1.37	1.25	1.40	0.17	0.46	0.69	1.18
24	0.77	0.86	0.06	1.97	0.15	0.59	0.13	1.90

Table 7.7: Thresholds for policies #5 - #8.

Hour	Policy # 9 On	Policy # 9 Off	Policy # 10 On	Policy # 10 Off	Policy # 11 On	Policy # 11 Off	Policy # 12 On	Policy # 12 Off
1	0.24	0.71	0.33	0.37	0.34	0.72	0.23	1.99
2	0.47	1.61	0.84	1.66	1.68	1.93	0.17	0.94
3	0.15	1.32	0.95	1.07	0.34	1.35	0.75	1.93
4	0.82	1.59	0.24	1.92	0.72	1.64	0.71	1.09
5	0.75	1.81	1.85	1.91	0.38	1.93	0.48	1.93
6	1.79	1.96	0.64	0.87	0.26	1.57	0.57	1.66
7	0.77	1.67	0.04	0.62	0.42	1.72	0.84	1.11
8	1.00	1.78	0.11	1.09	1.44	1.70	0.42	0.74
9	1.10	1.88	0.62	1.21	0.56	0.91	0.82	1.73
10	1.81	1.92	0.52	1.66	1.91	1.94	0.11	0.73
11	0.30	0.49	1.00	1.19	0.28	1.79	0.30	1.98
12	0.82	0.99	0.41	0.78	0.02	0.54	0.37	0.86
13	0.38	0.83	0.02	0.35	0.27	1.83	0.23	0.49
14	0.08	0.43	0.75	0.90	0.09	0.57	0.57	1.06
15	0.52	1.39	0.20	1.34	0.61	1.36	0.85	1.70
16	0.06	0.42	0.47	1.23	0.18	1.34	0.16	1.32
17	0.06	0.48	0.31	1.70	0.49	1.42	1.07	1.80
18	0.52	1.27	0.91	1.99	0.39	1.70	0.28	1.58
19	1.41	1.87	0.33	1.72	0.58	1.22	0.03	0.77
20	1.11	1.71	0.52	0.90	0.02	1.07	1.43	1.85
21	0.30	0.41	0.22	1.45	0.69	1.24	1.00	1.01
22	0.61	1.21	1.27	1.77	1.33	1.74	0.70	0.92
23	1.87	1.98	0.15	1.76	1.39	1.94	1.00	1.71
24	0.13	0.36	0.23	1.61	0.35	1.72	0.33	1.15

Table 7.8: Thresholds for policies #9 - #12.

Hour	Thresholds On	Thresholds Off
1	0.86	1.16
2	1.06	1.36
3	1.21	1.46
4	1.26	1.51
5	1.31	1.51
6	1.26	1.41
7	1.06	1.26
8	0.81	1.01
9	0.51	0.76
10	0.36	0.56
11	0.21	0.46
12	0.16	0.36
13	0.11	0.31
14	0.11	0.31
15	0.06	0.31
16	0.06	0.31
17	0.11	0.31
18	0.11	0.31
19	0.11	0.31
20	0.16	0.36
21	0.21	0.41
22	0.21	0.46
23	0.31	0.46
24	0.51	0.66

Table 7.9: Thresholds for optimal policy.

The backwards iteration algorithm uses an ordinal optimization approach at each stage, instead of over the entire policy space. It is necessary to select one and only one policy at each hour, as otherwise the number of policies would increase exponentially with the number of stages. To compare the two ordinal optimization approaches on a relatively even ground, we will sample 20 policies at each stage and run 50 Monte Carlo simulations per policy. With $N = 24$, this amount to calculating 24000 simulations, as compared to 25000 simulations for the first approach. (Note that in the backwards iteration algorithm, each simulation averages 12 hours instead of 24.) The results are given in Table 7.10 and Table 7.11. The expected profit is higher for this one policy than for any of the selected policies in the previous algorithm. The policy is also a better approximation of the optimal policy. Since the two sets of simulations for the on and off decisions are independent, the confidence level that "on" is actually optimal may be estimated. By the Central Limit Theorem [42], the averages of the cost for each simulation approach a normal distribution; for the given data, the probability that the "on" average is lower (more profitable) than "off" is 62.55%.

Using identical price samples for simulations at the same stage improves the results further. Although the simulated profit in Table 7.12 is slightly lower, the calculated policy in Table 7.13 more closely follows the optimal policy of Table 7.9. By measuring the difference between the "on" and "off" simulations, the power of using correlated simulations becomes clear. This difference has an average value of 12.76 over 50 simulations, with a standard deviation of 11.65. Since the standard deviation of the sample mean (12.76) is $11.65/\sqrt{50} = 1.65$, the probability that the average difference of the two simulations exceeds zero is 100.00%; i.e. using the policy in Table 7.13, it is virtually certain that leaving the generator on at 10 PM is the better choice. Using 50 policies per stage, with only 20 simulations per policy, gives a solution even closer to the optimum, as shown in Tables 7.14 and 7.15. The confidence level in the decision is 99.98%, derived from an expected difference of 13.79 with a standard deviation of 16.59.

The ordinal optimization algorithm uses much less computation than dynamic programming; the DP algorithm required 84.307s of user time, while the ordinal optimization algorithm using backwards iteration and independent simulations needed only 8.136s of user time, and 9.15s of elapsed time. Using identical price paths for all policies at a given stage not only further improves the results; it also requires even less computation. Simulating 50 polices per stage, with 20 simulations per policy, used only 1.114s of user time (2.49s of elapsed time) and gave a result with a very high level of confidence (99.98%).

Off		On		Optimal
Profit	Std. Dev.	Profit	Std. Dev.	Decision
348.98	29.25	364.85	36.63	On

Table 7.10: Results of backwards iteration using independent simulations of 20 policies per stage.

	Thresholds	
Hour	On	Off
1	0.41	1.36
2	1.87	1.99
3	1.28	1.70
4	1.52	1.95
5	1.14	1.66
6	0.10	0.85
7	0.58	1.13
8	0.64	1.67
9	0.08	0.96
10	0.90	1.85
11	0.19	0.38
12	0.05	0.36
13	0.42	1.78
14	0.41	0.43
15	0.17	0.22
16	0.19	0.61
17	0.05	1.89
18	0.13	0.72
19	0.08	0.52
20	0.05	0.98
21	0.00	0.23
22	0.50	1.98
23	0.52	0.81
24	0.17	0.78

Table 7.11: Estimated optimal policy from backwards iteration using independent simulations of 20 policies per stage.

Off		On		Optimal
Profit	Std. Dev.	Profit	Std. Dev.	Decision
342.87	30.59	355.64	30.68	On

Table 7.12: Results of backwards iteration using correlated simulations of 20 policies per stage.

	Thresholds	
Hour	On	Off
1	0.97	1.24
2	1.00	1.85
3	1.21	1.61
4	1.28	1.78
5	1.24	1.40
6	1.32	1.40
7	1.89	1.96
8	0.92	0.97
9	0.30	0.72
10	0.11	0.51
11	0.36	0.37
12	0.13	0.55
13	0.25	0.29
14	0.29	0.38
15	0.12	0.69
16	0.17	0.27
17	0.30	0.47
18	0.10	0.33
19	0.02	0.35
20	0.03	0.14
21	0.34	0.61
22	0.26	0.34
23	0.26	0.47
24	0.55	0.74

Table 7.13: Estimated optimal policy from backwards iteration using correlated simulations of 20 policies per stage.

	Off		On		Optimal
Profit	Std. Dev.	Profit	Std. Dev.	Decision
394.30	81.14	408.09	81.80	On

Table 7.14: Results of backwards iteration using correlated simulations of 50 policies per stage.

	Thresholds	
Hour	On	Off
1	0.82	1.03
2	1.08	1.99
3	1.43	1.57
4	1.07	1.81
5	1.46	1.70
6	1.56	2.00
7	1.02	1.78
8	0.69	1.05
9	0.56	0.91
10	0.22	0.39
11	0.08	0.66
12	0.07	0.17
13	0.12	0.23
14	0.26	0.49
15	0.12	0.48
16	0.07	0.07
17	0.02	0.35
18	0.20	0.26
19	0.25	0.27
20	0.05	0.33
21	0.26	0.34
22	0.08	0.50
23	0.37	0.46
24	0.55	0.74

Table 7.15: Estimated optimal policy from backwards iteration using correlated simulations of 50 policies per stage.

Given a numerical specification of the unit commitment problem for a power producer, an optimal solution may be calculated by computer, using both dynamic programming and ordinal optimization. The dynamic programming solution has a rigid theoretical foundation, although a discretization is needed to find a solution. The ordinal optimization method is subject to more chance, but a solution may be calculated quickly; this feature is particularly important when extensions of the unit commitment problem are considered in the following chapters. Note from Figure 7.2 that the top 5% of solutions have an expected profit of at least 200. The calculated solution from the backwards iteration using 50 policies per stage (Table 7.14) has an expected profit which lies 2.54 standard deviations above 200; this suggests that this solution is in the top 5% of all solutions with 99.4% confidence. With 20 policies per stage (Table 7.12), the expected profit lies 5.07 standard deviations above 200, suggesting that this is "good enough" with virtual certainty. A similar conclusion may be made for the top performing policies in Table 7.5. Since ordinal optimization performs well at finding "good enough" solutions for the basic unit commitment problem, we have confidence when applying it to more complicated formulations of unit commitment.

CHAPTER 8
FORWARD CONTRACTS AND FUTURES

Most market scenarios anticipate the development of a futures market. There are many possible forms for an electric futures market, including basic call and put options [7], forward contracts [33], callable forwards [34], and bid-based power pools [10]. Interruptible contracts [22], also known as recallable contracts or non-firm contracts, may be treated as callable forwards. This chapter will examine the impact of derivative securities on the profits of electric power producers and unit commitment strategies.

8.1 Forward Contracts

The deregulated electricity marketplace, whether poolco or bilateral, will likely include a market for forward contracts [7, 33, 34]. A forward contract is an agreement between a buyer and seller that a commodity (in this case electric power) will be delivered on a specified date in the future at a specified fixed price. The date specified by the contract is called the delivery date, while the price is known as the forward price. By contrast, the price of a unit power delivered immediately is referred to as the spot price. This type of contract offers both sides protection from possible future changes in price. In the electric marketplace, the contract may cover power delivery for an hour, a day, or a year or more.

In a forward contract marketplace, the price for a forward contract on a given date will vary over time, eventually converging to the spot price as the delivery date approaches. At each market time interval, which may be as frequent as one hour, the price of a forward contract will in general change. The buyer at that time can then sell the old contract and buy a new one at the new price. A buyer whose demand is inelastic to price can lock in the original forward contract price by maintaining the original demand level, while a buyer with more elastic demand can reduce consumption if forward prices rise and realize a profit on the sale of the original forward contract [33].

The forward price may be described as the market price for delivery of a commodity at a fixed time in the future. The forward price may also be interpreted as the risk-neutral expectation of the spot price at delivery. The forward price is a function of the expected price of the commodity on the delivery date. To show this, we first note that the present value at time t of an asset S at time T in the future is equal to the expected future value discounted at a risk-adjusted interest rate r_β:

$$PV_t\{S(T)\} = E\{S(T)\}e^{-r_\beta(T-t)} \tag{8.1}$$

using continuous compounding. In the case of a forward contract, since both buyer and seller have agreed to the contract, the contract has a present value of zero at the time of agreement, and consequently the present value of the spot price equals the present value of the forward price. The spot price is discounted at a rate that reflects the risk in price changes, while the forward price is discounted at a risk-free rate r_0. Therefore, if a forward market for a commodity exists, the forward price, discounted by the risk-free rate of return, is equal to the expected present value of the commodity on the delivery date [7]:

$$p_{t \to T} = E\{p(T)\}e^{(r_0-r_\beta)(T-t)} \tag{8.2}$$

Here $p_{t \to T}$ is the forward price at time t for delivery at time T and $p(T)$ is the price at time T, which may be treated as a random variable at time t. Since $r_\beta > r_0$, it follows that $p_{t \to T} < E\{p(T)\}$.

Futures are similar to forward contracts. Futures have prespecified terms, and they are traded on organized exchanges. Also, futures are typically "marked to market" every day; this means that the difference in the futures prices between the start and the end of the day is exchanged between the buyer and seller of a futures contract. The resulting payment schedule is spread out over the duration of the futures contract, instead of occurring as a lump settlement on the delivery date; although this difference in payment means that futures have a different value from the corresponding forward contract, this difference is often neglected in practice. A futures market for electricity, defined by delivery at the California-Oregon border (COB), started trading in March 1996 [5, 7].

8.2   Producer Profits with Forward Contracts

If at some earlier time, a power producer has agreed to a forward contract to sell P_{FC} units of power at the price p_f for delivery at time k, then when time k does roll around, the hourly profit is:

$$
\begin{aligned}
\pi_k &= u_k(p_k P_G - c_G(P_G) - I(x_k < 0)S + P_{FC}(p_f - p_k)) \\
&\quad + (1 - u_k)(P_{FC}(p_f - p_k) - c_f - I(x_k > 0)T)
\end{aligned}
\tag{8.3}
$$

If the generator is running, then the producer may generate P_G units of power. P_G is chosen to maximize the hourly profit, having knowledge of the spot price p_k. If P_G exceeds P_{FC}, then P_{FC} units are sold at the forward price p_f while the remaining units are sold at p_k. If P_G is less than P_{FC}, then the producer can still fulfill the terms of the forward agreement by selling the P_G units generated at the forward price and then purchasing the difference on the spot market at the spot price p_k. This principle also applies if the generator is off ($P_G = 0$); all P_{FC} units may be purchased at the spot price p_k and resold at the forward price p_f. The profit term $P_{FC}(p_f - p_k)$ in equation (8.3) accounts for all of these possibilities.

In order to fulfill delivery of a forward contract by purchasing from the spot market, it is necessary that spot market power be substitutable. While this is true in many cases (the end user simply draws power from a large network), it is not true if congestion is present, as not all generators may be able to supply the purchaser because of bottlenecks in the transmission grid. In these cases, the generator may have to purchase power at a price higher than the spot price p_k if extra power is needed to fulfill delivery terms in a forward contract; in extreme cases, the generator may be the only possible supplier on the entire grid, and therefore the generator is obliged to commit the generator to be turned on during hours where forward sales have been made. The generator can be relieved of the commitment to deliver in such cases only by buying back the forward contract, which likely would be very expensive.

On the other hand, many forward contracts and other derivative securities are settled financially rather than physically. In such cases, the two parties simply exchange money according to the forward and spot prices at the time of delivery. In a forward contract, if the spot price exceeds the forward price, then the seller pays the buyer the price difference for each unit sold in the forward contract; if the spot price is less than the forward price, the price difference times the quantity is paid by the buyer to the seller. Both parties may also conduct transactions on the spot market independently of each other. If no congestion exists, then the two methods of settlement have essentially identical outcomes for the profits of the generator and the consumer in the contract; however, as described above, the presence of grid congestion may cause the outcomes to differ. For now, we assume that there is no congestion present (or equivalently, that all forward contracts are settled financially).

Since the extra term in equation (8.3) due to forward contracts is not a function of the unit commitment decision u_k, the expected profit at hour k

may be rewritten as:

$$\pi_k = P_{FC}(p_f - p_k) + u_k(p_k P_G - c_G(P_G) - I(x_k < 0)S)$$
$$- (1 - u_k)(c_f + I(x_k > 0)T) \tag{8.4}$$

Since the forward contract profit is also independent of P_G, it is easy to see from equation (8.4) that the optimal generation level is the same as if forward contracts are not present; i.e. the marginal cost of generation is equal to the price.[8] Furthermore, since the unit commitment decision does not affect the forward contract profit term, this term may be pulled out of the minimization that is done when the optimal unit commitment strategy is selected. In other words, the unit commitment problem has no dependence on the forward contract commitment of the supplier.

8.3   Forward Contract Strategies

Given that a generator owner is able to sell power in forward markets for a given delivery date, the owner would like to know how much power should in fact be sold in the forward market. The optimal decision for this problem depends on two factors: the prices for forward contracts, and the desired optimization criterion. The owner may choose to simply minimize the variance of profit or to minimize a weighted sum of expected profit and variance. We will consider this problem for both a single hour of delivery and for contracts which require delivery over a period of many hours.

8.3.1   One Hour Forward Contracts

The total profit at hour k will depend on the unit commitment decision u_k:

$$\pi_k = P_{FC}(p_f - p_k) + u_k(p_k P_G - c_G(P_G)) - (1 - u_k)c_f \tag{8.5}$$

The startup and shutdown costs can be ignored, as they are constants. Substituting the optimal generation level:

$$P_G = \frac{p_{(MC)k} - b}{2a} \tag{8.6}$$

the profit is, for a quadratic cost curve:

$$\pi_k = P_{FC}(p_f - p_k) + u_k\left(\frac{p_k p_{(MC)k} - bp_k}{2a} - \frac{p_{(MC)k}^2 - b^2}{4a} - c\right)$$
$$- (1 - u_k)c_f \tag{8.7}$$

This may be rewritten as:

$$\pi_k = -\frac{u_k}{4a}p_{(MC)k}^2 + \frac{u_k}{2a}p_k p_{(MC)k} - \left(P_{FC} + u_k\frac{b}{2a}\right)p_k$$

$$+ P_{FC}p_f + u_k\left(\frac{b^2}{4a} - c\right) - (1 - u_k)c_f \qquad (8.8)$$

The profit is a quadratic function of the random variables p_k and $p_{(MC)k}$; for simplicity, we will denote:

$$\pi_k = C_{A1}p_{(MC)k}^2 + C_{A2}p_k p_{(MC)k} + C_B p_k + C_C \qquad (8.9)$$

The mean of the profit at hour k is:

$$\underset{p_k}{E}\{\pi_k\} = C_{A1}(\bar{p}_{(MC)k}^2 + \sigma_{(MC)k}^2) + C_{A2}\underset{p_k}{E}\{p_k p_{(MC)k}\} + C_B\bar{p}_k + C_C \qquad (8.10)$$

As before, $\bar{p}_k$ is the mean of p_k, $\bar{p}_{(MC)k}$ is the mean of $p_{(MC)k}$ (from equation (5.15)), and $\sigma_{(MC)k}^2$ is the variance of $p_{(MC)k}$, which is calculated by applying equation (5.18).

To find the variance of the profit, we first note that any sum of random variables X, Y, and Z has a variance of:

$$\text{var}(X + Y + Z) = \underset{X,Y,Z}{E}\{((X - \overline{X}) + (Y - \overline{Y}) + (Z - \overline{Z}))^2\} \qquad (8.11)$$

Expanding the square gives:

$$\text{var}(X + Y + Z) = \underset{X,Y,Z}{E}\{(X - \overline{X})^2 + (Y - \overline{Y})^2 + (Z - \overline{Z})^2$$
$$+ 2(X - \overline{X})(Y - \overline{Y})$$
$$+ 2(X - \overline{X})(Z - \overline{Z}) + 2(Y - \overline{Y})(Z - \overline{Z})\} \qquad (8.12)$$

This is equivalent to:

$$\text{var}(X + Y + Z) = \text{var}(X) + \text{var}(Y) + \text{var}(Z)$$
$$+ 2\text{cov}(X, Y) + 2\text{cov}(X, Z) + 2\text{cov}(Y, Z) \qquad (8.13)$$

Applying equation (8.13) to the profit of equation (8.9) gives:

$$\text{var}(\pi_k) = C_{A1}^2\text{var}(p_{(MC)k}^2) + C_{A2}^2\text{var}(p_k p_{(MC)k}) + C_B^2\text{var}(p_k)$$
$$+ 2C_{A1}C_{A2}\text{cov}(p_{(MC)k}^2, p_k p_{(MC)k}) + 2C_{A1}C_B\text{cov}(p_{(MC)k}^2, p_k)$$
$$+ 2C_{A2}C_B\text{cov}(p_k p_{(MC)k}, p_k) \qquad (8.14)$$

Note from equation (8.8) that the variance does not depend on the forward price p_f. Also note that the forward contract quantity P_{FC} affects only C_B

and C_C. The value of P_{FC} that minimizes the profit variance can therefore be found by differentiation:

$$\frac{\partial \mathrm{var}(\pi_k)}{\partial C_B} = 2C_B\mathrm{var}(p_k) + 2C_{A1}\mathrm{cov}(p^2_{(MC)k}, p_k) + 2C_{A2}\mathrm{cov}(p_k p_{(MC)k}, p_k) = 0$$

(8.15)

Solving for C_B:

$$C_B = -\frac{C_{A1}\mathrm{cov}(p^2_{(MC)k}, p_k) + C_{A2}\mathrm{cov}(p_k p_{(MC)k}, p_k)}{\mathrm{var}(p_k)}$$

(8.16)

From the standpoint of minimal variance, the optimal quantity of power to sell in forward contracts for hour k is:

$$P_{FC} = -\frac{C_{A1}\mathrm{cov}(p^2_{(MC)k}, p_k) + C_{A2}\mathrm{cov}(p_k p_{(MC)k}, p_k)}{\mathrm{var}(p_k)} - u_k\frac{b}{2a}$$

(8.17)

If the generator is off at time k, then both C_{A1} and C_{A2} are zero, and the variance is simply:

$$\mathrm{var}(\pi_k) = P^2_{FC}\mathrm{var}(p_k)$$

(8.18)

and therefore it is easy to see that minimal variance is achieved if $P_{FC} = 0$.

In order to determine the expected profit during a given hour in the distant future, we need to estimate the probability that the generator will be up at that time. This probability can often be assumed as zero or one in many situations (for example, a low-cost generator may be expected to run 24 hours, while many generators may be expected to run during peak hours); however, in some cases, it is more difficult to estimate. One approach would be to estimate the probability that the price exceeds a threshold level; however, it is necessary to account for the correlation between u_k and p_k.

This formulation also assumes that the cost curve is constant or known with certainty. Many investment problems often assume the existence of only one random input, as multiple random inputs greatly increase the complexity of the problem, and some random effects can be combined into a single random variable [7].

8.3.2   Multi-Hour Forward Contract

Many forward contracts and futures specify a single price for delivery of electricity over many hours. For example, the COB (California-Oregon border) peak futures contract specifies delivery for 16 hours a day on 5 days a week over an entire month [5]. The profit over all hours of a multi-hour forward contract (denoted $C_{T(FC)}$) from hour h_a to hour h_b is:

$$\pi_{T(FC)} = \sum_{k=h_a}^{h_b} \pi_k \tag{8.19}$$

Substituting equation (8.8) for π_k:

$$
\begin{aligned}
\pi_{T(FC)} = \ & -\frac{1}{4a} \sum_{k=h_a}^{h_b} u_k p_{(MC)k}^2 + \frac{1}{2a} \sum_{k=h_a}^{h_b} u_k p_k p_{(MC)k} + P_{FC} p_f (h_b - h_a) \\
& - P_{FC} \sum_{k=h_a}^{h_b} p_k - \frac{b}{2a} \sum_{k=h_a}^{h_b} u_k p_k + P_{FC} p_f (h_b - h_a) \\
& + \left(\frac{b^2}{4a} - c \right) \sum_{k=h_a}^{h_b} u_k - c_f \sum_{k=h_a}^{h_b} (1 - u_k)
\end{aligned} \tag{8.20}
$$

Here p_f is the forward price for one hour of electricity; the prices for all hours of the contract are identically p_f.

This problem is essentially the same as the one hour problem developed earlier; P_{FC} is chosen to minimize a desired criterion, such as variance of $\pi_{T(FC)}$. It is, however, more difficult to solve; while the mean is generally easy to calculate analytically (due to linearity of expectation), the variance of equation (8.20) is much harder to determine, as there are many correlated random variables whose covariances all need to be known. Monte Carlo simulations may be a practical approach to estimating these variances.

8.3.3 Example

To illustrate the use of forward contracts to hedge risk, a simple numerical example is given here. First, recall from the price process model that:

$$x_t = b_t + mL_t \tag{8.21}$$

where x_t is the logarithm of price, b_t is an intercept which follows a mean-reverting process, and L_t represents the demand, which can be forecasted. If time t is sufficiently far into the future, then the expected value of x_t is:

$$E\{x_t\} = \bar{b} + mE\{L_t\} \tag{8.22}$$

Since the process of b_t and the demand are assumed to be independent processes, the variance of x_t is:

$$\text{var}(x_t) = \text{var}(b_t) + \text{var}(mL_t) \tag{8.23}$$

Using the long-range variance of a mean-reverting process:

$$\text{var}(x_t) = \frac{\sigma_b^2}{1 - e^{-2\eta}} + m^2 \text{var}(L_t) \tag{8.24}$$

where σ_b^2 denotes the variance of b.

We will now use some representative numbers. From Chapter 6, the mean-reverting process of b_t has parameters $\bar{b} = 0.788$, $\eta = 0.317$, $m = 7.05 \times 10^{-5}$, and $\sigma_b^2 = 2.60 \times 10^{-2}$. On a Monday with a high of 65°F and a low of 45°F at Philadelphia, the expected load between 3:00 and 4:00 PM is approximately 25984. The error of this estimate is modeled as having a normal distribution with a standard deviation of 1600. Substituting these parameters into equations (8.22) and (8.24), $E\{x_t\} = 2.62$ and $\text{var}(x_t) = 6.81 \times 10^{-2}$. The price p_k is therefore lognormally distributed with mean 14.2 and standard deviation of 3.77 (from equations (6.16) and (6.17)).

The generator will be modeled with a cost curve $c_G(P_G) = P_G^2 + P_G + 9$ and generation limits $1 \leq P_G \leq 10$. To find the variances and covariances in equation (8.14), the truncated lognormal formulas in Appendix C may be used to find the expected values in the following equations:

$$\text{var}(X) = E(X^2) - [E(X)]^2 \tag{8.25}$$

$$\text{cov}(X, Y) = E(XY) - E(X)E(Y) \tag{8.26}$$

For the example data, the optimal value for P_{FC} from equation (8.17) turns out to be 7.21. If the price p_k is equal to its expected value (14.2), then the optimal quantity of power to produce is 6.6; this difference is a result of the quadratic cost function. The variance of π_k when $P_{FC} = 7.21$ is 21.67. If no forward contracts are issued ($P_{FC} = 0$), then the variance of π_k jumps to 761.90. The expected profit for this example is:

$$\underset{p_k}{E}\{\pi_k\} = 38.03 + P_{FC}(p_f - E\{p_k\}) \tag{8.27}$$

Forward contracts clearly provide a hedge against the risk of price uncertainty, although from equation (8.2) the forward price will offset the reduced risk by lowering the overall expected profit for stage k.

8.4   Temporal Forward Contract Problem

Up to this point, the question of choosing how much power to sell forward has been treated as a one-step problem; a quantity P_{FC} is selected, and then a random outcome is determined. In practice, forward prices evolve over a long period of time, providing many opportunities to change the amount of forward holdings. The problem of choosing the optimal forward contract amount at each time step for a fixed delivery date in the future may be formulated as a dynamic programming problem. In this formulation, the control variable is the quantity of forward contracts sold for delivery at time N, and the state is equal to the control of the previous time period; i.e. it is the amount of power currently sold forward for delivery at time N. Note that a generator owner would simultaneously solve this problem for all future delivery dates for which forward sales are possible.

At time N, the spot price p_N is known, and the total profit for the generator owner having sold x_N units forward is:

$$\pi_N = p_N(P_G - x_N) - c(P_G) \tag{8.28}$$

At time N, this quantity has zero variance. At time $N - 1$, $u_{N-1} = x_N$ units are sold through forward contracts at a forward price $p_{(N-1) \to N}$. The total profit for stages $N - 1$ and N is:

$$\pi_{N-1} + \pi_N = p_{(N-1) \to N}(u_{N-1} - x_{N-1}) + p_N(P_G - u_{N-1}) + c(P_G) \tag{8.29}$$

with variance:

$$\text{var}(\pi_{N-1} + \pi_N) = \text{var}(p_N(P_G - u_{N-1}) - c(P_G)) \tag{8.30}$$

Starting at time zero, the total profit for all stages is [33]:

$$\sum_{k=0}^{N} \pi_k = p_{0 \to N} u_0 + \sum_{k=1}^{N-1} p_{k \to N}(u_k - u_{k-1}) + p_N(P_G - u_{N-1}) - c(P_G) \tag{8.31}$$

The variance of this profit is:

$$\text{var}\left(\sum_{k=0}^{N} \pi_k\right) = \text{var}\left(\sum_{k=1}^{N-1} p_{k \to N}(u_k - u_{k-1}) + p_N(P_G - u_{N-1}) - c(P_G)\right) \tag{8.32}$$

In principle, this is a dynamic programming problem. The first major problem is determining an objective function. An obvious candidate would be to minimize:

$$-\sum_{k=0}^{N} \pi_k + W \text{var}\left(\sum_{k=0}^{N} \pi_k\right) \tag{8.33}$$

where W is a weighting factor. However, it is not clear that this choice of objective function is justified by finance theory. A practical choice would be to maximize the present value of the total profit:

$$PV\{Q\} = E\{Q\}e^{-\mu N} \tag{8.34}$$

This can be transformed by applying a logarithm:

$$\log PV\{Q\} = \log E\{Q\} - \mu N \tag{8.35}$$

This has the same form as the objective in equation (8.33); however, the relation between the variance and the risk-adjusted rate of return μ is not clear.

Using (8.33) as an objective, the temporal forward contract problem then may be solved by the dynamic programming recursion:

$$J_N(\mathbf{x}_N) = g_N(\mathbf{x}_N) \tag{8.36}$$

$$J_k(\mathbf{x}_k) = \min_{\mathbf{u}_k \in U_k(\mathbf{x}_k)} \underset{\mathbf{w}_k}{E} \left\{ g_k(\mathbf{x}_k, \mathbf{u}_k, \mathbf{w}_k) + J_{k+1}(\mathbf{u}_k) \right\} \tag{8.37}$$

where the cost functions g_k are:

$$g_N(x_N) = \underset{p_N}{E} \left\{ p_N(x_N - P_G) + c(P_G) \right\} \tag{8.38}$$

$$g_k(x_k) = p_{k \to N}(x_k - u_k) + W \mathrm{var} \left(\sum_{i=k}^{N} \pi_i \right) - W \mathrm{var} \left(\sum_{i=k+1}^{N} \pi_i \right) \tag{8.39}$$

This problem is theoretically solvable; however, it requires a large amount of price information, including the mean and variance of all future forward prices up to the delivery date and the expected variance of the spot price at future times. The profit function can also be amended to include both fixed and variable transaction costs. A complete study of this problem is left for future research.

CHAPTER 9

RESERVE MARKETS FOR POWER SYSTEM RELIABILITY

Up to this point, we have assumed that a power producer is always capable of producing the power that it sells. Unfortunately, in the real world, mechanical devices sometimes break down, and generators are no exception. It has been remarked that "it is not a question of whether or not a particular piece of equipment will fail, but rather when it will fail" [17]. It is widely observed in electricity restructuring debates that electricity is not storable, and consequently temporary production failures can not be covered by inventory, as is the case with most other commodities. Instead, it is necessary to have generation on the system operating at less than capacity, so that reserve power is readily available in case of a generator or line failure.

For a power producer, reliability poses two main questions. The first question is the provision of backup for the power that is sold to loads. The second question is to determine how much power should be sold on the spot market and how much generation should be held in reserve in order to maximize the expected profit. The exact nature of the provision of generation reserve is not clear at the time of this writing; however, we will use a generalized formulation that incorporates many possible forms. In particular, we will assume the existence of a market for reserve generation. We then consider the second question above (determination of optimal selling strategy for a power producer) first and return to the first question later in the chapter.

9.1 General Form of a Reserve Market

A market for reserve will operate concurrently with the spot market for power, although the reserve price p_R will be different from the spot price p. Whereas spot market power is sold and scheduled in advance of demand, reserve power must be available for immediate use in the event of unexpected contingencies, such as generator outages. Reserve power is a fundamentally different commodity from spot power. Like any other market, the reserve price reflects an equilibrium point between the supply and demand. The supply for reserve comes from generators, who also supply the spot market for electricity. The demand for reserve can come from any number of sources, depending on the exact nature of reliability maintenance in the market. An ISO may calculate and purchase all of the reserve needed for the system area in order to maintain a minimum standard. Alternatively, groups of generators may contract with each other to provide reserve for each other's transactions; in this case, a power producer is both supplying and demanding reserve. Loads may wish to buy reserve for their power. Reserve brokers may develop in the marketplace to purchase reserve power for their customers, who may be loads and/or generators.

The reserve price can either be higher than or lower than the spot price, depending on whether reserve payments are made for actual power delivered or for power that is merely reserved.

9.1.1 Payment for Power Delivered

In this scenario, a generator which sells power as reserve is paid the reserve price for that reserve only if the reserve power is actually used. The reserve price is therefore higher than the spot price, since excess generation capacity has a per unit cost that is higher than the spot price. (Generation with a lower marginal cost is sold for profit on the spot market.) In this case, a generator receives a profit on sales of reserve only for the time periods when the reserve actually needs to be generated. The generator receives zero profit if the reserve is not called.

9.1.2 Payment for Reserve Allocated

In this payment method, a generator receives the reserve price per unit of reserve power for every time period that the reserve is allocated and not used. If the reserve is used, then the generator receives the spot price for the reserve power that is generated. Since the reserve is not generated most of the time (hopefully!), reserve power has a very low expected cost, and hence the price of reserve will be much lower than the spot price of power. A generator receives a small profit for each time period in which the reserve is sold but not used; however, the generator will absorb a loss if the reserve is called. The reserve price p_R will be high enough such that the generator expects an overall long-term profit; otherwise, no reserve would be offered for sale.

9.1.3 Price Process for Reserve Price

Unfortunately, since there are no existing reserve markets, there is no empirical data available for building a price process model. However, it is clear that the reserve price and the spot price must be strongly correlated, since the quantity traded on the spot market largely defines the demand for reserve power. A hypothetical price model can be developed.

9.2 Individual Power Producer Strategies for Selling Reserve

If a power producer is able to sell power into a reserve market, then the producer's strategies for profit maximization in both the spot and reserve markets are intertwined. The producer decides to sell $P_{G(S)}$ in the spot market and $P_{G(R)}$ in the reserve market. The exact determination of $P_{G(S)}$ and $P_{G(R)}$ depends on the way reserve payments are made, although the results are very similar.

9.2.1 Payment for Power Delivered

For this payment method, $p < p_R$. During a given time period for a known price, the profit for a power producer is a random function with expectation:

$$
\begin{aligned}
E\{\pi_G\} = \ & pP_{G(S)} + rp_R(P_{G(T)} - P_{G(S)}) - (1 - r)(aP_{G(S)}^2 + bP_{G(S)} + c) \\
& - r(aP_{G(T)}^2 + bP_{G(T)} + c)
\end{aligned}
\tag{9.1}
$$

Here $P_{G(T)} = P_{G(S)} + P_{G(R)}$ and r is the probability that the reserve power is called and generated. An individual producer will choose $P_{G(S)}$ and $P_{G(R)}$ to maximize equation (9.1); these values may be found by differentiation:

$$
\frac{\partial E\{\pi_G\}}{\partial P_{G(S)}} = p - rp_R - (1 - r)(2aP_{G(S)} + b)
\tag{9.2}
$$

$$
\frac{\partial E\{\pi_G\}}{\partial P_{G(T)}} = rp_R - r(2aP_{G(T)} + b)
\tag{9.3}
$$

Setting these derivatives to zero, we have:

$$
\frac{p - rp_R}{1 - r} = 2aP_{G(S)} + b
\tag{9.4}
$$

$$
p_R = 2aP_{G(T)} + b
\tag{9.5}
$$

These equations are easy to interpret. Equation (9.4) indicates that power is sold on the spot market until the marginal cost of power equals an adjusted version of the spot price p. The adjustment reflects the fact that the marginal units of power have very little profit and would be more profitable on average if they are sold in the reserve market at the higher reserve price p_R. Since r is typically very small, the adjustment to p will also be very small. Equation (9.5) means that the power producer will sell reserve until the marginal cost reaches the price of reserve.

Both $P_{G(S)}$ and $P_{G(T)}$ must fall between the upper and lower generation limits. The optimal decision for these two variables with generation limits is essentially the same as the earlier model without a reserve market. Since the derivatives of profit are monotonically decreasing, if equation (9.4) yields a value of $P_{G(S)}$ that is less than P_G^{min}, the optimal choice is $P_{G(S)} = P_G^{min}$. Similarly, if $P_{G(S)}$ is calculated to be greater than P_G^{max}, then $P_{G(S)} = P_G^{max}$. The same is true for $P_{G(T)}$. Mathematically, this relationship may be written using the marginal cost limits defined in Chapter 5:

$$p_{MC}^{min} = 2aP_G^{min} + b \tag{9.6}$$

$$p_{MC}^{max} = 2aP_G^{max} + b \tag{9.7}$$

By further defining p_{eff}, the "effective" price for spot market sales:

$$p_{eff} = \frac{p - rp_R}{1 - r} \tag{9.8}$$

the prices p_{eff} and p_R may be written as truncated random variables:

$$P_{(MCeff)k} = \begin{cases} p_{MC}^{min} & p_{eff} \leq p_{MC}^{min} \\ p_{eff} & p_{MC}^{min} < p_{eff} < p_{MC}^{max} \\ p_{MC}^{max} & p_{eff} \geq p_{MC}^{max} \end{cases} \tag{9.9}$$

$$P_{(MCR)k} = \begin{cases} p_{MC}^{min} & p_R \leq p_{MC}^{min} \\ p_R & p_{MC}^{min} < p_R < p_{MC}^{max} \\ p_{MC}^{max} & p_R \geq p_{MC}^{max} \end{cases} \tag{9.10}$$

Using this notation, the optimal $P_{G(S)}$ and $P_{G(T)}$ are:

$$P_{G(S)} = \frac{P_{(MCeff)k} - b}{2a} \tag{9.11}$$

$$P_{G(T)} = \frac{P_{(MCR)k} - b}{2a} \tag{9.12}$$

The expected profit in stage k is given by:

$$\begin{aligned} E\{\pi_G\} = {} & (1 - r)(p_{eff}P_{G(s)} - (aP_{G(S)}^2 + bP_{G(S)} + c)) \\ & + r(p_R p_{G(T)} - (aP_{G(T)}^2 + bP_{G(T)} + c)) \end{aligned} \tag{9.13}$$

9.2.2 Payment for Reserve Allocated

Under this payment method, $p \gg p_R$. The expected profit of an individual generator is determined by using the same procedure as in the last section:

$$E\{\pi_G\} = pP_{G(S)} + ((1-r)p_R + rp)(P_{G(T)} - P_{G(S)})$$
$$- (1-r)(aP_{G(S)}^2 + bP_{G(S)} + c) - r(aP_{G(T)}^2 + bP_{G(T)} + c) \quad (9.14)$$

As before, we find the maximum of equation (9.14) by differentiation:

$$\frac{\partial E\{\pi_G\}}{\partial P_{G(S)}} = p - ((1-r)p_R + rp) - (1-r)(2aP_{G(S)} + b) \quad (9.15)$$

$$\frac{\partial E\{\pi_G\}}{\partial P_{G(T)}} = ((1-r)p_R + rp) - r(2aP_{G(T)} + b) \quad (9.16)$$

The zero point of the derivatives gives the strategy for maximizing profit:

$$p - p_R = 2aP_{G(S)} + b \quad (9.17)$$

$$p + (r^{-1} - 1)p_R = 2aP_{G(T)} + b \quad (9.18)$$

Equation (9.17) indicates that the marginal cost of power sold on the spot market should be reduced by the reserve price, in comparison to the original formulation without a reserve market. One disadvantage of this payment method can be observed from the optimal strategy for a power producer given by equation (9.18): the amount of reserve offered for sale is highly sensitive to r, because of the dependence on r^{-1}.

Generation limits are handled by the same procedure used in the preceding section. First, we define the following "effective" prices, which determine the optimal marginal cost:

$$p_{eff(S)} = p - p_R \quad (9.19)$$

$$p_{eff(R)} = p + (r^{-1} - 1)p_R \quad (9.20)$$

The corresponding truncated random variables, using the limits of equations (9.6) and (9.7), are:

$$P_{(MCeff(S))k} = \begin{cases} p_{MC}^{min} & p_{eff(S)} \leq p_{MC}^{min} \\ p_{eff(S)} & p_{MC}^{min} < p_{eff(S)} < p_{MC}^{max} \\ p_{MC}^{max} & p_{eff(S)} \geq p_{MC}^{max} \end{cases} \quad (9.21)$$

$$P_{(MCeff(T))k} = \begin{cases} p_{MC}^{min} & p_{eff(T)} \leq p_{MC}^{min} \\ p_{eff(T)} & p_{MC}^{min} < p_{eff(T)} < p_{MC}^{max} \\ p_{MC}^{max} & p_{eff(T)} \geq p_{MC}^{max} \end{cases} \quad (9.22)$$

$P_{G(S)}$ and $P_{G(T)}$ then become, for optimal profit:

$$P_{G(S)} = \frac{p_{(MCeff(S))k} - b}{2a} \tag{9.23}$$

$$P_{G(T)} = \frac{p_{(MCeff(T))k} - b}{2a} \tag{9.24}$$

The expected profit in stage k is given by:

$$\begin{aligned}
E\{\pi_G\} &= (1-r)(p_{eff(S)}P_{G(s)} - (aP_{G(S)}^2 + bP_{G(S)} + c)) \\
&\quad + r(p_{eff(T)}p_{G(T)} - (aP_{G(T)}^2 + bP_{G(T)} + c)) \tag{9.25}
\end{aligned}$$

9.3   Provision of Reserve for Transactions

We now return to the first question regarding reserve posed earlier: How is reserve provided for power sold on the open market from generators to loads? The responsibility for provision of generation reserve can rest either with the loads or with the generators, although the end result (price paid by loads) will likely be the same in either case. Reserve can also be provided by the ISO as a system service, with its cost included in the charge for system use. If the load is responsible for reserve, the load has flexibility to precisely determine a desired tradeoff between reliability and price. A group of loads may collectively purchase a block of reserve under a joint agreement; purchases through a reserve broker have a similar net result. A load may also choose to be fully or partially interruptible and thus avoid or reduce the cost of reserve.

If generators are responsible for providing reserve, then they can form collective agreements in a similar fashion to loads as described above, either through negotiation or through a reserve broker. The price paid by loads for power will be somewhat higher than if loads are responsible for reserve, with the difference reflecting the cost of reserve. Loads that choose interruptible power will pay a lower price.

The development of an area-wide market for reserve has an advantage of offering a lower price than if bilateral reserve agreements are made. This concept is best illustrated by an example in which reserve payments are for power that is generated. If generator A has a 500 MW contract and needs backup for this contract, then in order to induce generator B to sell reserve, A would need to offer a reserve price equal to B's marginal generation cost at the generation level of both spot power for B's sales and all 500 MW of A's reserve power. However, in a reserve market, the 500 MW reserve can be spread across all generators in the area, which means that B might only offer 100 MW in the reserve market. In this case, the marginal cost of generating the 100 MW of reserve is clearly less than the cost of generating all 500 MW of reserve, leading to a lower price for reserve. If the total reserve offered among all generators in the area exceeds the largest amount of power generated by any one generator, then the $(N-1)$ contingency criterion will be satisfied, assuming that transmission constraints are not a factor.

9.4 Effect of Reserve Market on Unit Commitment

The inclusion of a reserve market has two principal effects on the unit commitment algorithm. First, the ability to sell reserve power affects the profit maximization strategy, as shown earlier in section 9.2, and therefore the expected one-stage profit is also changed. The correlation between reserve calls and prices may need to be included in the expected profit; a higher price may imply a higher value of r. Second, the responsibility for reserve and the possibility of generator failure (a major reason for having reserve in the first place) should also be included in the expected profit. Note that both factors do not change the available unit commitment options; therefore, the unit commitment algorithm can be modified to account for the reserve market by adjusting the expected one-stage profit, and if necessary, adding another continuous state variable, which is the price of reserve.

Since reserve market sales are added to the problem formulation, the reliability of the producer's own generator should also be considered. If a single generator, with failure probability f, experiences a failure and is unable to produce the power it sold, it must buy that power from the reserve market at the reserve price p_R (assuming that reserve payments are for power delivered). In this case, the expected profit per stage should be adjusted according to:

$$\underset{a_G}{E}\{\pi_G\} = (1-f)E\{\pi_G\} + f(p-p_R)P_{G(S)} \qquad (9.26)$$

The optimal $P_{G(S)}$ also needs to be adjusted, according to:

$$\frac{\partial \underset{a_G}{E}\{\pi_G\}}{\partial P_{G(S)}} = (1-f)\frac{\partial E\{\pi_G\}}{\partial P_{G(S)}} + f(p-p_R) = 0 \qquad (9.27)$$

This equation may be transformed to:

$$\frac{\partial E\{\pi_G\}}{\partial P_{G(S)}} = \frac{f(p_R - p)}{1-f} \qquad (9.28)$$

If f is small, then the effects of a producer's generator failure are also small and may be neglected. The exact form of equation (9.26) for a specific situation may have many possibilities, depending on the form of the reserve market.

9.5 Example

The following hypothetical example, derived from the example in Chapter 7, illustrates a possible reserve market unit commitment strategy. The reserve price will be modeled as:

$$\ln p_R = K_R + e_R + \ln p_k \tag{9.29}$$

K_R is a constant, while e_R is normally distributed with zero mean; for this numerical example, $K_R = 0.7$ and $\mathrm{var}(e_R) = 0.0625$. Reserve payments are made for power delivered. The probability of reserve calls will be taken as $r = 5 \times 10^{-3}$, independent of the price. The failure probability of the generator is $f = 1 \times 10^{-4}$. To simplify the problem, the generator will use the strategy $p_{eff} \approx p_k$, since p_{eff} is not strictly lognormal. With this strategy, the expected profit per stage is:

$$
\begin{aligned}
\underset{p_k}{E}\{\pi_k\} \;=\; & u_k \Bigg((1-f)\Bigg[(1-r)\underset{p_k}{E}\left\{ \frac{p_k p_{(MC)k} - b p_k}{2a} - \frac{p_{(MC)k}^2 - b^2}{4a} - c \right\} \\
& + r\underset{p_k}{E}\left\{ \frac{p_R p_{(MCR)k} - b p_R}{2a} - \frac{p_{(MCR)k}^2 - b^2}{4a} - c \right\} \Bigg] \\
& + f\underset{p_k}{E}\left\{ (p_R - p_k)\frac{p_{(MC)k} - b}{2a} \right\} - I(x_k < 0)S \Bigg) \\
& - (1-u_k)(c_f + I(x_k > 0)T)
\end{aligned}
\tag{9.30}
$$

The first two expectations may be evaluated by using the truncated lognormal distributions in Chapter 5. The last expectation requires an evaluation of $\underset{p_k}{E}\{p_R p_{(MC)k}\}$. Taking an exponential of both sides of equation (9.29):

$$p_R = p_k e^{K_R} e^{e_R} \tag{9.31}$$

Therefore:

$$\underset{p_k}{E}\{p_R p_{(MC)k}\} = \underset{p_k}{E}\{p_k p_{(MC)k} e^{K_R} e^{e_R}\} \tag{9.32}$$

e^{K_R} is a constant and may be factored out of the expectation. Furthermore, e^{e_R} is independent of p_k (and hence $p_{(MC)k}$) and therefore:

$$\underset{p_k}{E}\{p_R p_{(MC)k}\} = e^{K_R} \underset{p_k}{E}\{p_k p_{(MC)k}\} E\{e^{e_R}\} \tag{9.33}$$

since the expectation of a product of uncorrelated random variables is the product of the expectations. The expected profit therefore becomes:

$$
\underset{p_k}{E}\{\pi_k\} = u_k \Bigg((1-f)\Bigg[(1-r)\underset{p_k}{E}\left\{ \frac{p_k p_{(MC)k} - b p_k}{2a} - \frac{p_{(MC)k}^2 - b^2}{4a} - c \right\}
$$

Status	Expected Profit	Optimal Decision
On since 9:00 PM	411.52	On
On since 8:00 PM	416.50	On
On since 7:00 PM	416.50	On
Off since 9:00 PM	404.62	Off
Off since 8:00 PM	404.62	Off

Table 9.1: Optimal decision and expected profit for each state at 10:00 PM

Status	Expected Profit	Optimal Decision
On since 10:00 PM	387.95	On
On since 9:00 PM	401.36	On
On since 8:00 PM	403.92	Off
Off since 10:00 PM	407.92	Off
Off since 9:00 PM	407.92	Off

Table 9.2: Optimal decision and expected profit for each state at 11:00 PM

$$
\begin{aligned}
&+ r \underset{p_k}{E} \left\{ \frac{p_R p_{(MCR)k} - b p_R}{2a} - \frac{p_{(MCR)k}^2 - b^2}{4a} - c \right\} \Bigg] \\
&+ f \underset{p_k}{E} \left\{ \frac{b(p_k - p_R) - p_k p_{(MC)k}(1 - e^{K_R} E\{e^{e_R}\})}{2a} \right\} \\
&\quad - I(x_k < 0)S \Bigg) - (1 - u_k)(c_f + I(x_k > 0)T) \qquad (9.34)
\end{aligned}
$$

Using the same cost and price process data as in Chapter 7, the results of the unit commitment algorithm for 10:00 PM and 11:00 PM are given in Tables 9.1 and 9.2. The optimal unit commitment decisions are the same as in Chapter 7, but the opportunity for sales of reserve power somewhat increases the expected profit.

CHAPTER 10
UNIT COMMITMENT IN CONGESTED
TRANSMISSION SYSTEMS

The presence of physical limitations on the amount of power that may be transferred through a given transmission line gives rise to a very vigorous debate on how to account for these limits in a deregulated utility environment. Many pricing schemes that account for line flow limits are being discussed and debated [10, 11, 23, 35], but no consensus has been reached. This uncertainty makes it difficult to predict the effects of congestion on a power producer. Congestion will likely change the prices of power as well as reserve, but maximum generation limits at certain locations may also be implemented. If no such limits are present, then the strategy for selling power and unit commitment is essentially as described in the previous chapters, with the only difference being that the price for power reflects the presence of congestion, and the price process must model the price that the generator actually receives. If congestion does impose limits on power generation for a given producer, then a more complicated model is needed.

10.1 Probabilistic Model of Congestion

As with reserve, we will develop a generalized formulation for modeling the effects of congestion on individual power producers.

10.1.1   Modifications to Price Model

If price data is available from a congested system, then a price model may be developed directly from the data. Such a price model would account for line congestion. One possible form for a congested model takes the price model for uncongested systems ($p_{k(uncong)}$) and includes a congestion penalty $p_{k(cpen)}$ as a second random input [36]:

$$p_k = p_{k(uncong)} - p_{k(cpen)} \tag{10.1}$$

The price p_k from this model is used for the optimization algorithms. It is important to note that $p_{k(uncong)}$ and $p_{k(cpen)}$ will generally be correlated.

10.1.2   Congestion Model with Limits

This model assumes that there is a maximum limit P_{Clim} of power that can be sold. Like P_G^{max}, P_{Clim} is an upper limit on production. However, P_G^{max} is a fixed physical quantity, whereas P_{Clim} varies from one time period to the next. P_{Clim} is treated as a random variable; additionally, it follows an auto-regressive process similar to the price process model of Chapter 6. As with the reserve price model, the limit due to congestion is likely correlated with the price of power in addition to the amount of congestion during the previous hour.

10.2   Producer Strategy under Congestion

We will now determine the optimal selling strategy for a power producer in both the power and reserve markets when confronted with the possibility of congestion. Since P_{Clim} is a maximum generation limit, the optimal strategy for a seller is the same as in Chapter 9, with P_G^{max} replaced by P_{Clim} if the congestion limit is lower than the maximum generation limit. Furthermore, this formulation assumes that any power that cannot be sold due to congestion may still be offered for sale on the reserve market. The probability r that the reserve is used takes into account the congestion constraints on the system; if reserve generation is needed, then it is used from generators that do not cause any line flow limits to be exceeded.

Given the preceding formulation, we can adjust the expected cost per stage to reflect the probability of congestion. Using the notation from equation (9.26):

$$\underset{a_G}{E}\{\pi_G\} = (1 - f)E\{\pi_G\} + f(p - p_R)P_{G(S)} \tag{10.2}$$

the expected profit, accounting for congestion, may be written as a double integral:

$$\underset{P_{Clim},p_k}{E}\{\pi_G\} = \int_0^\infty \int_0^\infty \text{dens}(P_{Clim} = P', p_k = p'')$$
$$\times \underset{a_G}{E}\{\pi_G | P_G^{max} = P', p_k = p''\}dP'dp'' \qquad (10.3)$$

since p_k and P_{Clim} are in general correlated. The double integral may also be reduced to single integrals over the conditional expectation of π_G:

$$\underset{P_{Clim},p_k}{E}\{\pi_G\} = \int_0^\infty \text{dens}(p_k = p'') \underset{a_G,P_{Clim}}{E}\{\pi_G | p_k = p''\}dp'' \qquad (10.4)$$

$$\underset{P_{Clim},p_k}{E}\{\pi_G\} = \int_0^\infty \text{dens}(P_{Clim} = P') \underset{a_G,p_k}{E}\{\pi_G | P_G^{max} = P'\}dP' \qquad (10.5)$$

The congestion limit may be modeled as a continuous variable or as a variable taking only certain discrete values. In the latter case, the integrals in equations (10.3) and (10.5) are replaced with summations over each possible value of P_{Clim}.

Unfortunately, the preceding equations are difficult to use for the general case where p_k and P_{Clim} are correlated. Typically, P_{Clim} will have a probability distribution as a function of the price p_k. Such a specification makes it easy to evaluate the expectation inside the integral of equation (10.4), but performing the integration will be messy. The integration of equation (10.5) is generally easy to perform, but the expectation requires knowledge of both the expectation and variance of p_k *conditioned* on $P_{Clim} = P'$; these conditional statistics are not in general equal to the overall mean of p_k.

10.3 Example

To illustrate a hypothetical congestion situation, let P_{Clim} be independent of price and have a random distribution of:

$$\text{Prob}(P_{Clim} = P) = \begin{cases} 0.1 & : P = 5 \\ 0.1 & : P = 7 \\ 0.8 & : P = \infty \end{cases} \qquad (10.6)$$

For this example, we neglect the effects of the reserve market. Applying equation (10.5), the cost per stage is:

$$\underset{p_k}{E}\{\pi_k\} = u_k\left(0.8\underset{p_k}{E}\left\{\frac{p_k P_{(MC)k} - bp_k}{2a} - \frac{p^2_{(MC)k} - b^2}{4a} - c\right\}\right.$$
$$\left. + 0.1\underset{p_k}{E}\left\{\frac{p_k P_{(MCa)k} - bp_k}{2a} - \frac{p^2_{(MCa)k} - b^2}{4a} - c\right\}\right.$$

Status	Expected Profit	Optimal Decision
On since 9:00 PM	393.81	On
On since 8:00 PM	399.02	On
On since 7:00 PM	399.02	On
Off since 9:00 PM	387.90	Off
Off since 8:00 PM	387.90	Off

Table 10.1: Optimal decision and expected profit for each state at 10:00 PM

Status	Expected Profit	Optimal Decision
On since 10:00 PM	370.26	On
On since 9:00 PM	383.98	On
On since 8:00 PM	386.98	Off
Off since 10:00 PM	390.98	Off
Off since 9:00 PM	390.98	Off

Table 10.2: Optimal decision and expected profit for each state at 11:00 PM

$$+ 0.1 \underset{p_k}{E} \left\{ \frac{\frac{p_k p_{(MCb)k} - b p_k}{2a} - p^2_{(MCb)k} - b^2}{4a} - c \right\} - I(x_k < 0)S \right)$$
$$- (1 - u_k)(c_f + I(x_k > 0)T) \tag{10.7}$$

$p_{(MCa)k}$ is a truncated normal variable with an upper limit of $15 = 2a \times 7 + b$; $p_{(MCb)}$ is a truncated normal variable with an upper limit of $11 = 2a \times 5 + b$.

The results of the unit commitment algorithm for 10:00 PM and 11:00 PM with the preceding congestion model applied to the example of Chapter 7 are given in Tables 10.1 and 10.2. As in the reserve market example, the anticipated congestion limits do not affect the unit commitment decision, although they do reduce the expected profit.

10.4 Solution of Unit Commitment under Congestion and Reserve

In principle, the solution of unit commitment does not change significantly in the presence of reserve markets or congested lines; however, a number of practical problems do arise. First, actual price data from reserve markets and congested systems is needed to develop useful price models for unit commitment. Methods such as those described in Chapter 6 may be used to create the price models. These models depend on the stochastic behavior of all of the generator cost curves and all of the load demand curves for market participants; hence, they are very difficult to model without empirical data. Second, the price models for a congested system with a reserve market will in general have three continuous state variables instead of just one, except in certain specialized cases (such as the examples in the last two chapters). If there are N_S continuous state variables, and each is discretized into N_d values, then the total number of states is:

$$(t_{up} + t_{dn})^{N_G} N_d^{N_S}$$

Clearly, the number of states grows exponentially with the number of state variables, unless the discretization of each variable is greatly reduced. Third, owners of multiple generators have the option of using their own generators to provide reserve. While it is not clear whether this is more cost-effective than selling that reserve on the reserve market, this option, along with congestion considerations, may require that all generators be optimized simultaneously instead of individually; doing so means that exponential growth is encountered with respect to the number of generators N_G. To obtain near-optimal solutions in these cases, ordinal optimization and Monte Carlo methods will likely be needed.

CHAPTER 11
CONCLUSIONS

This book focuses on problems requiring zero-one decisions (i.e. on-off) for which price is an important factor. The objective of the problem is to maximize the expected profit or benefit over a time horizon of many hours, days, or even weeks. Such problems are faced by electric generator owners as well as large industrial users. Although this book presents examples with only two control options, the techniques are generalizable to problems where several options may be considered. Since the profit during any hour in the future is a random quantity when viewed from the present, the problem is stochastic in nature.

Throughout this book, the unit commitment problem is developed in detail as a representative example of a price-based decision problem. Unit commitment is the process by which generating units are turned on and off. Unit commitment decisions are subject to many constraints, such as minimum up and down times. Unit commitment is usually solved over a time horizon ranging from 24 hours to a week. This book focuses on the unit commitment problem in transition from a regulated environment to one that is deregulated and market-driven. However, the concepts in this book are not limited to unit commitment and may be applied to a much broader class of decision problems.

This book examines the role of unit commitment in the newly forming electric power market. The analysis follows three major steps. First, the unit commitment problem is posed for an individual producer in a deregulated market. A market model is developed to perform price forecasting. Finally, the problem is solved using a method that accounts for inter-temporal effects. This book presents dynamic programming [1] as an essential tool for making decisions. Dynamic programming (DP) software is developed to illustrate algorithms on hypothetical market architectures. The unit commitment problem is also extended to include forward contracts, reserve markets, and congestion limits.

This book develops several new approaches to unit commitment. A broad definition of unit commitment for the present utility structure is shown. The simplifications needed to apply this broad model to practical unit commitment situations are illustrated; in particular, the stochastic nature of the problem is often reduced to one which is deterministic. Most dynamic optimization [3] is currently done using a deterministic model, which treats future values of price and demand as known. A deterministic solution is in general suboptimal, sometimes significantly so, when applied to the corresponding stochastic problem [1].

The unit commitment problem is then defined for an individual power producer in a deregulated industry structure. Many features of the broad model of unit commitment in a regulated industry are included in the deregulated formulation of the unit commitment problem; notably, a stochastic framework is maintained. The deregulated problem has many fewer decision options, allowing for the inclusion of many more features. A major contribution of this book is to carry out the optimization by assuming a stochastic model. In this book, quantities such as price and demand are treated as random variables with known mean and variance, and the optimization aims to maximize the expected value of the objective function.

Generators cannot produce an unlimited amount of power; there is an upper and lower limit on the power which is produced. Since the expected profit at a given hour depends on the generation level, truncated random variables are defined and introduced in order to account for generation limits. These variables are treated as being distributed continuously between two limits but also having positive probability of being equal to either limit. The optimal generation level, which is a function of a random price input, is well described by truncated random variables.

Unit commitment under deregulation requires a stochastic model of the price of electricity. Although a basic random walk or mean-reverting price model may be used to describe the price data, these models treat all 24 hours of the day equally and ignore changes in electricity demand at different times of the day. A representative model based on a time-varying mean is derived here from limited data using load estimates. The load estimates may themselves be obtained from the date of the year, estimates of economic growth, and the expected outdoor temperature; a particularly strong correlation between temperature and load usage was discovered from PJM pool data [27]. While the variance of load estimation reduces the usefulness of its application to price prediction, it still forms a good basis for changing demand patterns at different times of the day. As the US electricity market forms and matures, the accuracy of such models will improve as more information becomes available.

A direct solution by dynamic programming on a hypothetical numerical example is presented for optimal decision-making. An alternative solution method based on ordinal optimization is also presented and shown to require much less computation to obtain the same solution. Although the dynamic programming solution is feasible for the unit commitment model in this paper, additions to the price process model, consideration of market power (requiring all generators to be optimized simultaneously), and other possible changes will likely make a direct dynamic programming solution impractical; Monte Carlo solution methods using far less computation can account for many additional problem features while still giving an optimal or near-optimal solution with high probability. Because of the success of ordinal optimization techniques on a relatively small example, there is a great degree of confidence in their application to larger unit commitment problems.

This book includes forward contracts and futures, which are agreements to sell a fixed quantity of a commodity on a specified date in the future at a price fixed at the inception of the contract. Futures markets for electricity are in their infancy; however, independent power producers can use forward markets to greatly reduce the risk of future profits. Forwards and futures have a payoff which increases as the price drops. A more general optimization problem for futures can be defined which calculates the optimal forward contract position over time as spot and forward prices for electricity evolve; however, this problem is very complicated to solve.

This book examines various means for the provision of reserve generation in deregulated electricity markets. It is shown that reserve may be bought and sold, and it is a separate commodity from power in the spot market. As the number of generators participating in the reserve market increases in a given area, the price of reserve in that area drops. Interruptible loads have the same net effect as reserve generation; they provide a means of maintaining the balance of supply and demand in the event of a failure in the system. An example is presented illustrating how independent power producers make decisions when having the option of selling in both the spot power market and the reserve market. The price of reserve at equilibrium will balance the supply and demand of reserve, thus setting the most economical quantity of backup generation for the system.

Although a detailed formulation of the effects of congestion on power producers is not available at the time of this writing, an overview of the likely effects are considered. There are two main possibilities: a penalty factor on the price of power sold, and hard limits on power sales. A simple numerical example of sales limits is shown to illustrate how their effects on profits may be computed by power sellers; not surprisingly, the presence of congestion lowers profits.

There are many avenues of future research. The application of gaming theory is important to properly model both market power and pool bidding strategies. Further developments of the models in this book, particularly for congestion, are needed, particularly as more data becomes available, and the resulting increase in complexity will require improvements in solution methods. The temporal forward contract problem, using a present value objective criterion, is of great interest to power sellers. Also, several variables in this book were modeled as uncorrelated; an examination of the correlation of such variables may lead to different and more realistic results.

CALCULATION OF PARAMETERS FOR PRICE PROCESS

A.1 Exponential of a Normally Distributed Variable

Given a normally distributed random variable X with mean μ and variance σ^2, the expectation of e^X may be written:

$$E(e^X) = \frac{1}{\sigma\sqrt{2\pi}} \int_{-\infty}^{\infty} e^x e^{-\frac{(x-\mu)^2}{2\sigma^2}} dx \tag{A.1}$$

The exponents inside the integral may be combined:

$$E(e^X) = \frac{1}{\sigma\sqrt{2\pi}} \int_{-\infty}^{\infty} e^{-\frac{x^2 - 2\mu x + \mu^2 - 2\sigma^2 x}{2\sigma^2}} dx \tag{A.2}$$

By completing the square in the x^2 and x terms, this equation may be written:

$$E(e^X) = \frac{1}{\sigma\sqrt{2\pi}} \int_{-\infty}^{\infty} e^{-\frac{(x-(\mu+\sigma^2))^2 + \mu^2 - (\mu+\sigma^2)^2}{2\sigma^2}} dx \tag{A.3}$$

The exponential may be written as a product of two exponentials:

$$E(e^X) = e^{\mu + \frac{1}{2}\sigma^2} \int_{-\infty}^{\infty} \frac{e^{-\frac{(x-(\mu+\sigma^2))^2}{2\sigma^2}}}{\sigma\sqrt{2\pi}} dx \tag{A.4}$$

The integrand in equation (A.4) is simply the density function of a normal distribution of mean $\mu + \sigma^2$ and variance σ^2; therefore, its integral is 1.

The expected value of an exponential of a normally distributed variable is therefore:

$$E(e^X) = e^{\mu + \frac{1}{2}\sigma^2} \tag{A.5}$$

The derivation for the variance of e^X is very similar. Using the normal distribution density function:

$$\text{var}(e^X) = \frac{1}{\sigma\sqrt{2\pi}} \int_{-\infty}^{\infty} (e^x - e^{\mu + \frac{1}{2}\sigma^2})^2 e^{-\frac{(x-\mu)^2}{2\sigma^2}} dx \tag{A.6}$$

Upon expanding the square, the integral becomes three separate integrals:

$$\text{var}(e^X) = \frac{1}{\sigma\sqrt{2\pi}} \left[\int_{-\infty}^{\infty} e^{2x} e^{-\frac{(x-\mu)^2}{2\sigma^2}} dx - 2e^{\mu + \frac{1}{2}\sigma^2} \int_{-\infty}^{\infty} e^x e^{-\frac{(x-\mu)^2}{2\sigma^2}} dx \right.$$
$$\left. + e^{2\mu + \sigma^2} \int_{-\infty}^{\infty} e^{-\frac{(x-\mu)^2}{2\sigma^2}} dx \right] \tag{A.7}$$

The third integral is simply an integral of a probability density function, while the second integral is the mean of e^X. After combining the exponentials in the first integral, this equation becomes:

$$\text{var}(e^X) = \frac{1}{\sigma\sqrt{2\pi}} \int_{-\infty}^{\infty} e^{-\frac{x^2 - 2\mu x + \mu^2 + 4\sigma^2 x}{2\sigma^2}} dx - 2e^{2\mu + \sigma^2} + e^{2\mu + \sigma^2} \tag{A.8}$$

As before, we can complete the square in the exponential:

$$\text{var}(e^X) = \frac{1}{\sigma\sqrt{2\pi}} \int_{-\infty}^{\infty} e^{-\frac{(x-(\mu+2\sigma^2))^2 + \mu^2 - (\mu+2\sigma^2)^2}{2\sigma^2}} dx - e^{2\mu + \sigma^2} \tag{A.9}$$

After separating the integrand into two exponentials:

$$\text{var}(e^X) = e^{2\mu + 2\sigma^2} \int_{-\infty}^{\infty} \frac{e^{-\frac{(x-(\mu+2\sigma^2))^2}{2\sigma^2}}}{\sigma\sqrt{2\pi}} dx - e^{2\mu + \sigma^2} \tag{A.10}$$

The integral is one; therefore, the variance of an exponential of a normally distributed random variable is:

$$\text{var}(e^X) = e^{2\mu + \sigma^2}(e^{\sigma^2} - 1) \tag{A.11}$$

A.2   Nonlinear Least Squares Regression

The nonlinear regression problem for the mean-reverting intercept model may be written as:

$$Y = \beta_0 + \beta_1 X_1 + \beta_2 X_2 + \beta_1 \beta_2 X_3 \tag{A.12}$$

The product term in the unknown parameters makes this a nonlinear regression problem. There have been several recent publications regarding the topic of nonlinear regression methods [37, 38]. Solving a generalized nonlinear problem may be quite computationally expensive; however, the regression in equation (A.12) may be solved by using Newton's method for nonlinear systems. The residual sum of squares (the function to be minimized) may be written:

$$RSS = \sum_{i=1}^{n}(y_i - \beta_0 - \beta_1 x_{i1} - \beta_2 x_{i2} - \beta_1 \beta_2 x_{i3})^2 \tag{A.13}$$

The partial derivatives of RSS with respect to the β_i are:

$$\frac{\partial RSS}{\partial \beta_0} = \sum_{i=1}^{n} -2(y_i - \beta_0 - \beta_1 x_{i1} - \beta_2 x_{i2} - \beta_1 \beta_2 x_{i3}) \tag{A.14}$$

$$\frac{\partial RSS}{\partial \beta_1} = \sum_{i=1}^{n} -2(y_i - \beta_0 - \beta_1 x_{i1} - \beta_2 x_{i2} - \beta_1 \beta_2 x_{i3})(x_{i1} + \beta_2 x_{i3}) \tag{A.15}$$

$$\frac{\partial RSS}{\partial \beta_2} = \sum_{i=1}^{n} -2(y_i - \beta_0 - \beta_1 x_{i1} - \beta_2 x_{i2} - \beta_1 \beta_2 x_{i3})(x_{i2} + \beta_1 x_{i3}) \tag{A.16}$$

The least squares estimator is the point at which all three derivatives are zero. To save space, the following definitions will be used:

$$Sx_k = \sum_{i=1}^{n} x_{ik} \quad Sx_k^2 = \sum_{i=1}^{n} x_{ik}^2 \quad Sx_k x_m = \sum_{i=1}^{n} x_{ik} x_{im}$$

$$Sy = \sum_{i=1}^{n} y_i \quad Sx_k y = \sum_{i=1}^{n} x_{ik} y_i$$

All of these quantities are functions of the data, and not of the unknown parameters. Using this notation, the residual sum of squares is minimized when:

$$Sy - \beta_0 n - \beta_1 Sx_1 - \beta_2 Sx_2 - \beta_1 \beta_2 Sx_3 = 0 \tag{A.17}$$

$$
\begin{aligned}
Sx_1 y \quad - \quad & \beta_0 Sx_1 - \beta_1 Sx_1^2 - \beta_2(Sx_1 x_2 - Sx_3 y) \\
& - \beta_0 \beta_2 Sx_3 - 2\beta_1 \beta_2 Sx_1 x_3 - \beta_2^2 Sx_2 x_3 - \beta_1 \beta_2^2 Sx_3^2 = 0
\end{aligned} \tag{A.18}
$$

$$
\begin{aligned}
Sx_2 y \quad - \quad & \beta_0 Sx_2 - \beta_1(Sx_1 x_2 - Sx_3 y) - \beta_2 Sx_2^2 \\
& - \beta_0 \beta_1 Sx_3 - 2\beta_1 \beta_2 Sx_2 x_3 - \beta_1^2 Sx_1 x_3 - \beta_1^2 \beta_2 Sx_3^2 = 0
\end{aligned} \tag{A.19}
$$

This system can be solved using Newton's method. Starting from a guess $\mathbf{b}_{(0)}$, an improved guess $\mathbf{b}_{(1)}$ is obtained by [39]:

$$\mathbf{b}_{(k+1)} = \mathbf{b}_{(k)} - \mathbf{J}^{-1}(\mathbf{b}_{(k)})\mathbf{f}(\mathbf{b}_{(k)}) \tag{A.20}$$

$\mathbf{f} = \mathbf{0}$ is the system of equations to be solved; $\mathbf{J}$ is the Jacobian of $\mathbf{f}$. For equations (A.17) to (A.19), the element in row i, column j of $\mathbf{J}$ is $J_{i,j}$, given by:

$$J_{1,1} = -n \tag{A.21}$$

$$J_{2,2} = -Sx_1^2 - 2\beta_2 Sx_1 x_3 - \beta_2^2 Sx_3^2 \tag{A.22}$$

$$J_{3,3} = -Sx_2^2 - 2\beta_1 Sx_2 x_3 - \beta_1^2 Sx_3^2 \tag{A.23}$$

$$J_{1,2} = J_{2,1} = -Sx_1 - \beta_2 Sx_3 \tag{A.24}$$

$$J_{1,3} = J_{3,1} = -Sx_2 - \beta_1 Sx_3 \tag{A.25}$$

$$J_{2,3} = J_{3,2} = -(Sx_1 x_2 - Sx_3 y) - \beta_0 Sx_3 - 2\beta_2 Sx_2 x_3 - 2\beta_1 Sx_1 x_3 - 2\beta_1 \beta_2 Sx_3^2 \tag{A.26}$$

Note that $\mathbf{J}$ is symmetric. The iteration of equation (A.20) is repeated until the difference between two consecutive values of $\mathbf{b}_{(k)}$ is less than a chosen tolerance.

If the residuals are weighted by weighting factors w_i:

$$RSS = \sum_{i=1}^{n} w_i(y_i - \beta_0 - \beta_1 x_{i1} - \beta_2 x_{i2} - \beta_1 \beta_2 x_{i3})^2 \tag{A.27}$$

the solution may be found by modifying the definitions:

$$Sx_k = \sum_{i=1}^{n} w_i x_{ik} \quad Sx_k^2 = \sum_{i=1}^{n} w_i x_{ik}^2 \quad Sx_k x_m = \sum_{i=1}^{n} w_i x_{ik} x_{im}$$

$$Sy = \sum_{i=1}^{n} w_i y_i \quad Sx_k y = \sum_{i=1}^{n} w_i x_{ik} y_i$$

Using these definitions in the preceding equations and Jacobian formula will produce the weighted least-squares solution. Note that the weighting factors for the mean-reverting intercept model in Chapter 6 are:

$$w_i = f^{n-i} \tag{A.28}$$

A.3 Recursive Least Squares Algorithm

The goal of the recursive least squares algorithm is to take sequential data and update a least squares parameter estimate with a minimum of computation. Suppose that k sets of data have been received; the least squares estimate from this data will be denoted by $\beta_{(k)}$, which minimizes:

$$RSS_k = \sum_{i=1}^{k} f^{k-i}(y_i - \mathbf{x}_i^T \beta_{(k)})^2 \qquad (A.29)$$

y_i and $\mathbf{x}_i^T$ represent row i of the Y and X matrices respectively, while $0 < f \leq 1$ is a fading or "forgetting" factor. The solution for $\beta_{(k)}$ is derived from linear regression theory:

$$\left(\sum_{i=1}^{k} f^{k-i} \mathbf{x}_i \mathbf{x}_i^T \right) \beta_{(k)} = \sum_{i=1}^{k} f^{k-i} \mathbf{x}_i y_i \qquad (A.30)$$

If new data y_{k+1} and $\mathbf{x}_{k+1}$ are received, the updated estimate $\beta_{(k+1)}$ minimizes:

$$RSS_{k+1} = \sum_{i=1}^{k+1} f^{k-i}(y_i - \mathbf{x}_i^T \beta_{(k+1)})^2 \qquad (A.31)$$

and satisfies:

$$\left(\sum_{i=1}^{k+1} f^{k-i} \mathbf{x}_i \mathbf{x}_i^T \right) \beta_{(k+1)} = \sum_{i=1}^{k+1} f^{k-i} \mathbf{x}_i y_i \qquad (A.32)$$

Multiplying both sides by f:

$$\left(\sum_{i=1}^{k+1} f^{k+1-i} \mathbf{x}_i \mathbf{x}_i^T \right) \beta_{(k+1)} = \sum_{i=1}^{k+1} f^{k+1-i} \mathbf{x}_i y_i \qquad (A.33)$$

We can now define the matrix $\mathbf{Q}_k$ as:

$$\mathbf{Q}_k = \sum_{i=1}^{k} f^{k-i} \mathbf{x}_i \mathbf{x}_i^T \qquad (A.34)$$

Notice that $\mathbf{Q}_{k+1}$ may be written as a recursive function:

$$\mathbf{Q}_{k+1} = f\mathbf{Q}_k + \mathbf{x}_{k+1}\mathbf{x}_{k+1}^T \qquad (A.35)$$

Substituting for $\mathbf{Q}_{k+1}$ in equation (A.33):

$$\mathbf{Q}_{k+1}\beta_{(k+1)} = f \sum_{i=1}^{k} f^{k-i} \mathbf{x}_i y_i + \mathbf{x}_{k+1} y_{k+1} \qquad (A.36)$$

From equation (A.30), the summation in this equation is $\mathbf{Q}_k \beta_{(k)}$, giving:

$$\mathbf{Q}_{k+1}\beta_{(k+1)} = (f\mathbf{Q}_k + \mathbf{x}_{k+1}\mathbf{x}_{k+1}^T)\beta_{(k)} + \mathbf{x}_{k+1}y_{k+1} - \mathbf{x}_{k+1}\mathbf{x}_{k+1}^T\beta_{(k)} \qquad \text{(A.37)}$$

The quantity inside the parentheses is $\mathbf{Q}_{k+1}$; after multiplying both sides by $\mathbf{Q}_{k+1}^{-1}$:

$$\beta_{(k+1)} = \beta_{(k)} + \mathbf{Q}_{k+1}^{-1}\mathbf{x}_{k+1}(y_{k+1} - \mathbf{x}_{k+1}^T\beta_{(k)}) \qquad \text{(A.38)}$$

$$\mathbf{Q}_{k+1} = f\mathbf{Q}_k + \mathbf{x}_{k+1}\mathbf{x}_{k+1}^T \qquad \text{(A.39)}$$

Equations (A.38) and (A.39) are used to update the least squares estimate when data is received sequentially. The preceding derivation is from [40].

APPENDIX B
RESULTS OF REGRESSION OF LOAD vs. TEMPERATURE

This appendix gives coefficients for the cubic polynomial used to represent the load as a function of temperature. Residual plots and a list of possible outliers, for which the Studentized residual (denoted r_s) exceeds ± 3, are also given. A separate set of coefficients and plots is calculated for each hour of the day.

B.1 Regression on Same Day High Temperature

This section gives the coefficients and residuals for the regression results using the same day's high and low temperature, denoted as T_{hi} and T_{lo} respectively. The first four tables give the calculated regression coefficients along with a 95% confidence interval (indicated as "Upper" and "Lower"). Table B.5 gives R^2 and the F statistic along with estimates of the variance and standard deviation. The following tables include a plot of the Studentized residuals and a list of data points with large residuals for each hour.

Hour	Coefficient of T_{lo}^3			Coefficient of T_{lo}^2		
	Value	Lower	Upper	Value	Lower	Upper
9 AM - 10 AM	0.1163	0.0979	0.1347	−9.66	−12.15	−7.16
10 AM - 11 AM	0.1070	0.0883	0.1256	−8.39	−10.92	−5.86
11 AM - 12 PM	0.1008	0.0814	0.1201	−7.54	−10.17	−4.92
12 PM - 1 PM	0.0952	0.0753	0.1151	−6.85	−9.54	−4.15
1 PM - 2 PM	0.0910	0.0702	0.1117	−6.33	−9.14	−3.52
2 PM - 3 PM	0.0845	0.0629	0.1061	−5.59	−8.52	−2.67
3 PM - 4 PM	0.0799	0.0575	0.1023	−5.09	−8.13	−2.06
4 PM - 5 PM	0.0806	0.0565	0.1047	−5.30	−8.57	−2.03
5 PM - 6 PM	0.0895	0.0610	0.1180	−6.18	−10.04	−2.32
6 PM - 7 PM	0.0990	0.0718	0.1263	−7.02	−10.71	−3.33
7 PM - 8 PM	0.0932	0.0688	0.1177	−6.56	−9.88	−3.25
8 PM - 9 PM	0.0769	0.0568	0.0971	−4.91	−7.64	−2.18
9 PM - 10 PM	0.0696	0.0506	0.0886	−3.93	−6.51	−1.36
10 PM - 11 PM	0.0654	0.0470	0.0839	−3.30	−5.80	−0.80
11 PM - 12 AM	0.0622	0.0446	0.0799	−2.84	−5.23	−0.44

Table B.1: Regression results for T_{lo}^3 and T_{lo}^2.

Hour	Coefficient of T_{lo}			Constant Term		
	Value	Lower	Upper	Value	Lower	Upper
9 AM - 10 AM	93.87	−12.03	199.77	37623	35747	39500
10 AM - 11 AM	73.89	−33.59	181.36	36941	35036	38845
11 AM - 12 PM	64.76	−46.69	176.21	36070	34095	38044
12 PM - 1 PM	56.83	−57.76	171.43	35120	33090	37151
1 PM - 2 PM	50.16	−69.34	169.66	34473	32355	36591
2 PM - 3 PM	34.54	−89.86	158.95	33602	31397	35806
3 PM - 4 PM	24.70	−104.35	153.75	32984	30697	35271
4 PM - 5 PM	34.89	−103.92	173.70	33005	30545	35464
5 PM - 6 PM	28.35	−135.81	192.52	33783	30874	36692
6 PM - 7 PM	−4.70	−161.55	152.15	32718	29939	35498
7 PM - 8 PM	−24.76	−165.64	116.12	32478	29981	34974
8 PM - 9 PM	−67.26	−183.23	48.71	32755	30700	34810
9 PM - 10 PM	−98.47	−208.06	11.11	31815	29873	33757
10 PM - 11 PM	−130.89	−237.21	−24.57	29851	27967	31734
11 PM - 12 AM	−158.88	−260.66	−57.11	28178	26374	29981

Table B.2: Regression results for T_{lo} and the constant term.

Hour	Coefficient of T_{hi}^3			Coefficient of T_{hi}^2		
	Value	Lower	Upper	Value	Lower	Upper
9 AM - 10 AM	0.0362	0.0219	0.0504	−3.90	−6.53	−1.27
10 AM - 11 AM	0.0504	0.0360	0.0649	−5.82	−8.49	−3.15
11 AM - 12 PM	0.0623	0.0473	0.0773	−7.34	−10.11	−4.57
12 PM - 1 PM	0.0720	0.0566	0.0874	−8.54	−11.39	−5.69
1 PM - 2 PM	0.0788	0.0628	0.0949	−9.32	−12.29	−6.35
2 PM - 3 PM	0.0873	0.0706	0.1040	−10.42	−13.51	−7.33
3 PM - 4 PM	0.0957	0.0783	0.1130	−11.59	−14.80	−8.39
4 PM - 5 PM	0.1068	0.0881	0.1255	−13.33	−16.77	−9.88
5 PM - 6 PM	0.1219	0.0998	0.1440	−16.06	−20.14	−11.98
6 PM - 7 PM	0.1242	0.1032	0.1453	−17.28	−21.18	−13.39
7 PM - 8 PM	0.1111	0.0922	0.1301	−15.60	−19.10	−12.10
8 PM - 9 PM	0.1020	0.0864	0.1176	−14.05	−16.93	−11.17
9 PM - 10 PM	0.1007	0.0860	0.1154	−13.74	−16.46	−11.01
10 PM - 11 PM	0.0977	0.0834	0.1120	−13.57	−16.21	−10.93
11 PM - 12 AM	0.0897	0.0760	0.1034	−12.62	−15.15	−10.09

Table B.3: Regression results for T_{hi}^3 and T_{hi}^2.

Hour	Coefficient of T_{lo}		
	Value	Lower	Upper
9 AM - 10 AM	0.57	−154.98	156.12
10 AM - 11 AM	65.89	−91.97	223.74
11 AM - 12 PM	113.45	−50.25	277.14
12 PM - 1 PM	151.40	−16.91	319.72
1 PM - 2 PM	173.85	−1.68	349.37
2 PM - 3 PM	213.97	31.24	396.70
3 PM - 4 PM	258.38	68.83	447.92
4 PM - 5 PM	324.66	120.77	528.55
5 PM - 6 PM	453.64	212.52	694.76
6 PM - 7 PM	578.79	348.40	809.17
7 PM - 8 PM	538.66	331.74	745.59
8 PM - 9 PM	473.83	303.49	644.16
9 PM - 10 PM	459.23	298.27	620.18
10 PM - 11 PM	469.37	313.21	625.53
11 PM - 12 AM	444.32	294.84	593.81

Table B.4: Regression results for T_{hi}.

Hour	R^2	F	$\hat{\sigma}^2$	Standard Error
9 AM - 10 AM	0.8717	1021	1392592	1180.08
10 AM - 11 AM	0.8836	1140	1434206	1197.58
11 AM - 12 PM	0.8938	1264	1542260	1241.88
12 PM - 1 PM	0.9032	1402	1630642	1276.97
1 PM - 2 PM	0.9074	1471	1773250	1331.63
2 PM - 3 PM	0.9095	1509	1921802	1386.29
3 PM - 4 PM	0.9084	1490	2067946	1438.04
4 PM - 5 PM	0.8951	1282	2392733	1546.85
5 PM - 6 PM	0.8550	885	3346377	1829.31
6 PM - 7 PM	0.8548	884	3054958	1747.84
7 PM - 8 PM	0.8575	904	2464466	1569.86
8 PM - 9 PM	0.8852	1158	1670008	1292.29
9 PM - 10 PM	0.8971	1309	1491106	1221.11
10 PM - 11 PM	0.8960	1294	1403568	1184.72
11 PM - 12 AM	0.8939	1265	1286157	1134.09

Table B.5: Regression statistics and estimates of variance.

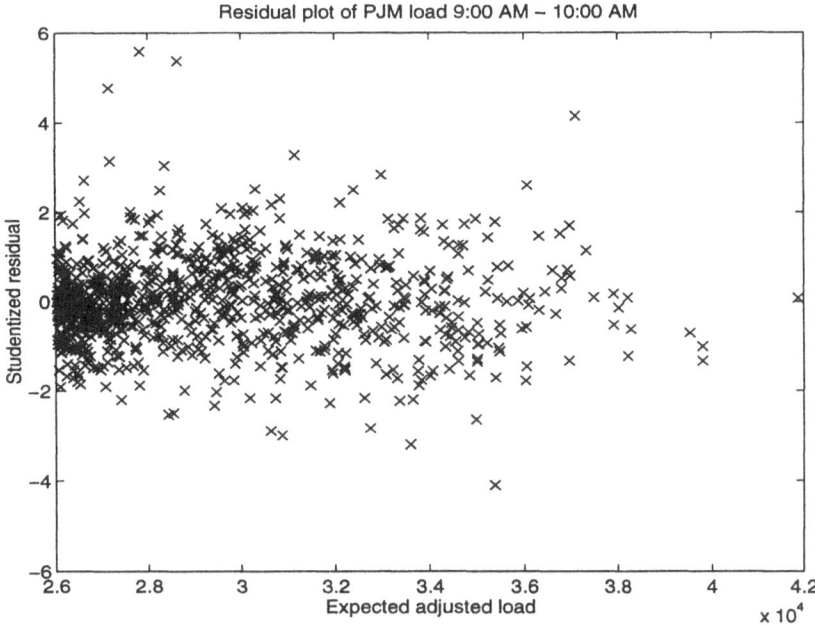

Figure B.1: Residual plot for 9:00 AM - 10:00 AM.

Studentized Residual	Day		Date	Adjusted Load Actual	Expected	High	Low
4.148	16	Sun	1/16/94	41805	37105	15	4
3.044	28	Fri	1/28/94	31919	28355	58	33
5.575	217	Fri	8/5/94	34272	27823	86	60
3.274	226	Sun	8/14/94	34993	31160	90	69
4.765	241	Mon	8/29/94	32699	27151	79	60
3.145	528	Mon	6/12/95	30864	27180	75	61
−3.186	725	Tue	12/26/95	29877	33600	31	22
−4.100	738	Mon	1/8/96	30609	35378	26	14
5.365	872	Tue	5/21/96	34796	28625	91	60

Table B.6: List of cases with high residuals, 9:00 AM - 10:00 AM.

r_s	Day		Date	Prev. Day Low	High	Low	High	Next Day Low	High
4.148	16	Sun	1/16/94	6	17	4	15	14	34
3.044	28	Fri	1/28/94	11	33	33	58	34	42
5.575	217	Fri	8/5/94	76	91	60	86	55	77
3.274	226	Sun	8/14/94	75	92	69	90	62	75
4.765	241	Mon	8/29/94	70	89	60	79	59	79
3.145	528	Mon	6/12/95	66	89	61	75	62	74
−3.186	725	Tue	12/26/95	27	32	22	31	21	31
−4.100	738	Mon	1/8/96	12	22	14	26	14	29
5.365	872	Tue	5/21/96	68	94	60	91	55	80

Table B.7: List of cases with high residuals, 9:00 AM - 10:00 AM.

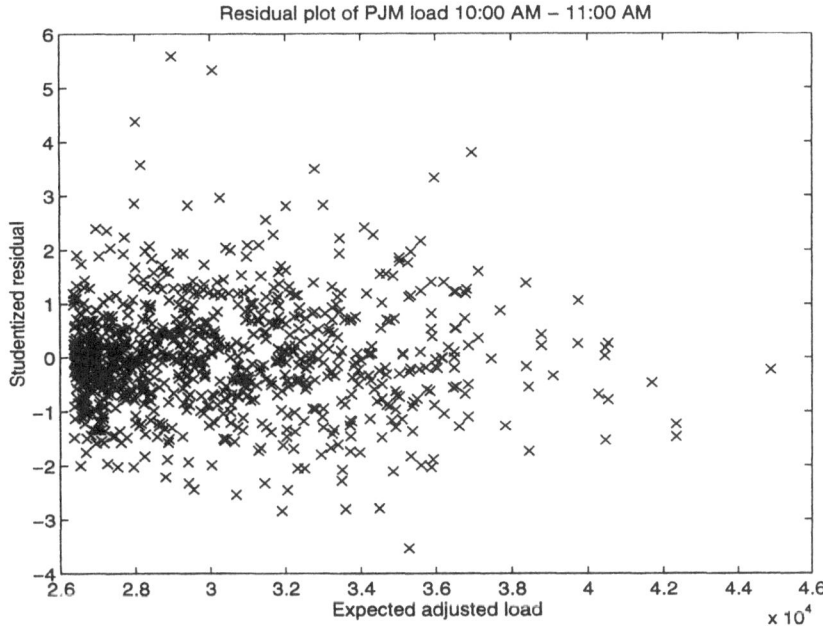

Figure B.2: Residual plot for 10:00 AM - 11:00 AM.

Studentized Residual	Day		Date	Adjusted Load Actual	Expected	High	Low
3.805	16	Sun	1/16/94	41339	36957	15	4
3.579	28	Fri	1/28/94	32407	28162	58	33
5.585	217	Fri	8/5/94	35523	28967	86	60
3.497	226	Sun	8/14/94	36935	32784	90	69
4.388	241	Mon	8/29/94	33213	28018	79	60
3.331	737	Sun	1/7/96	39894	35972	22	12
−3.533	738	Mon	1/8/96	31107	35287	26	14
5.332	872	Tue	5/21/96	36295	30070	91	60

Table B.8: List of cases with high residuals, 10:00 AM - 11:00 AM.

r_s	Day		Date	Prev. Day				Next Day	
				Low	High	Low	High	Low	High
3.805	16	Sun	1/16/94	6	17	4	15	14	34
3.579	28	Fri	1/28/94	11	33	33	58	34	42
5.585	217	Fri	8/5/94	76	91	60	86	55	77
3.497	226	Sun	8/14/94	75	92	69	90	62	75
4.388	241	Mon	8/29/94	70	89	60	79	59	79
3.331	737	Sun	1/7/96	10	20	12	22	14	26
−3.533	738	Mon	1/8/96	12	22	14	26	14	29
5.332	872	Tue	5/21/96	68	94	60	91	55	80

Table B.9: List of cases with high residuals, 10:00 AM - 11:00 AM.

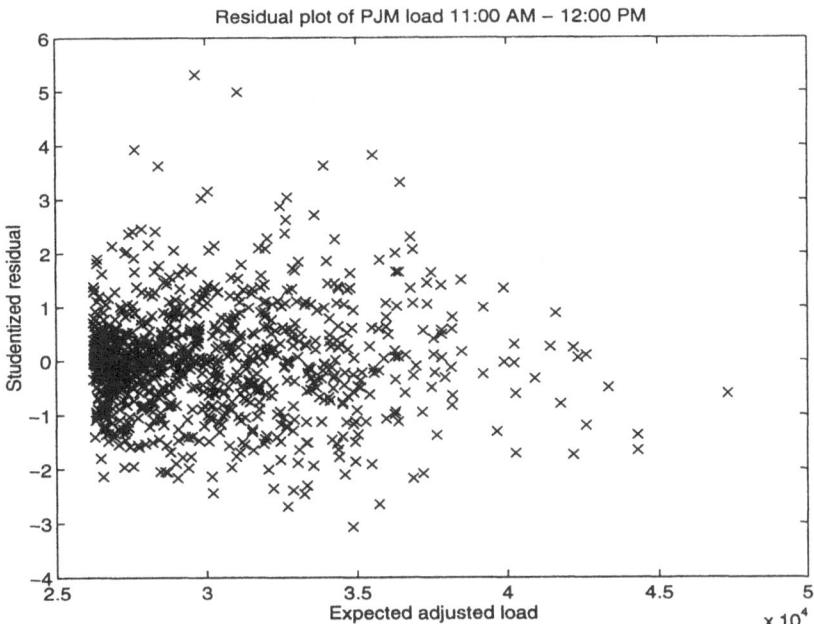

Figure B.3: Residual plot for 11:00 AM - 12:00 PM.

Studentized Residual	Day		Date	Adjusted Load Actual	Adjusted Load Expected	High	Low
3.314	16	Sun	1/16/94	40441	36475	15	4
3.929	28	Fri	1/28/94	32486	27660	58	33
3.029	44	Sun	2/13/94	33603	29864	45	30
3.039	213	Mon	8/1/94	36437	32709	80	71
5.305	217	Fri	8/5/94	36136	29668	86	60
3.620	226	Sun	8/14/94	38363	33908	90	69
3.619	241	Mon	8/29/94	32898	28440	79	60
3.822	737	Sun	1/7/96	40197	35542	22	12
−3.068	738	Mon	1/8/96	31087	34858	26	14
4.984	872	Tue	5/21/96	37109	31062	91	60
3.153	906	Mon	6/24/96	33967	30077	84	63

Table B.10: List of cases with high residuals, 11:00 AM - 12:00 PM.

Results of Regression of Load vs. Temperature

r_s	Day		Date	Prev. Day				Next Day	
				Low	High	Low	High	Low	High
3.314	16	Sun	1/16/94	6	17	4	15	14	34
3.929	28	Fri	1/28/94	11	33	33	58	34	42
3.029	44	Sun	2/13/94	25	33	30	45	24	35
3.039	213	Mon	8/1/94	72	84	71	80	74	88
5.305	217	Fri	8/5/94	76	91	60	86	55	77
3.620	226	Sun	8/14/94	75	92	69	90	62	75
3.619	241	Mon	8/29/94	70	89	60	79	59	79
3.822	737	Sun	1/7/96	10	20	12	22	14	26
−3.068	738	Mon	1/8/96	12	22	14	26	14	29
4.984	872	Tue	5/21/96	68	94	60	91	55	80
3.153	906	Mon	6/24/96	68	81	63	84	68	87

Table B.11: List of cases with high residuals, 11:00 AM - 12:00 PM.

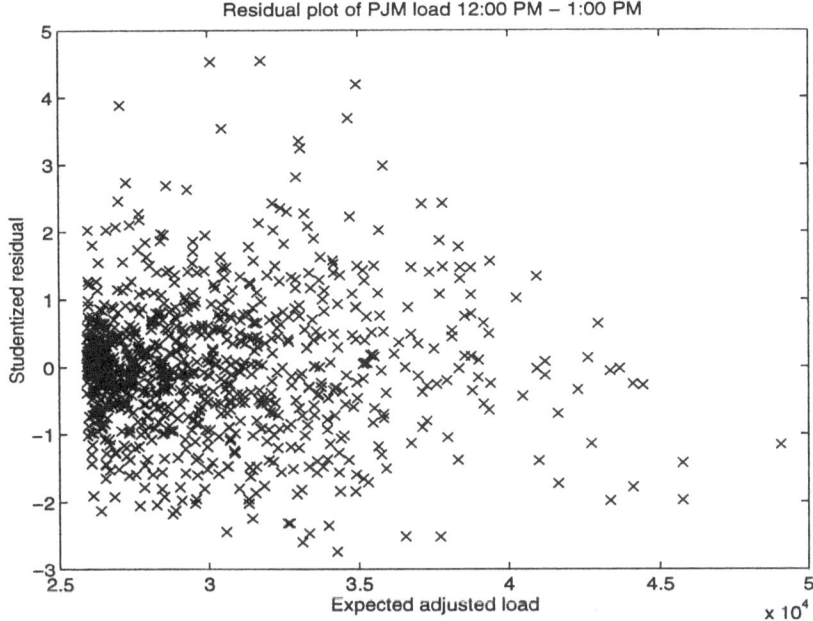

Figure B.4: Residual plot for 12:00 PM - 1:00 PM.

Studentized Residual	Day		Date	Adjusted Load Actual	Adjusted Load Expected	High	Low
3.885	28	Fri	1/28/94	31974	27067	58	33
3.353	213	Mon	8/1/94	37264	33039	80	71
4.531	217	Fri	8/5/94	35812	30108	86	60
3.693	226	Sun	8/14/94	39338	34666	90	69
4.193	737	Sun	1/7/96	40191	34946	22	12
4.545	872	Tue	5/21/96	37452	31770	91	60
3.245	899	Mon	6/17/96	37203	33089	88	67
3.541	906	Mon	6/24/96	34958	30472	84	63

Table B.12: List of cases with high residuals, 12:00 PM - 1:00 PM.

r_s	Day	Date		Prev. Day Low	High	Low	High	Next Day Low	High
3.885	28	Fri	1/28/94	11	33	33	58	34	42
3.353	213	Mon	8/1/94	72	84	71	80	74	88
4.531	217	Fri	8/5/94	76	91	60	86	55	77
3.693	226	Sun	8/14/94	75	92	69	90	62	75
4.193	737	Sun	1/7/96	10	20	12	22	14	26
4.545	872	Tue	5/21/96	68	94	60	91	55	80
3.245	899	Mon	6/17/96	66	91	67	88	70	82
3.541	906	Mon	6/24/96	68	81	63	84	68	87

Table B.13: List of cases with high residuals, 12:00 PM - 1:00 PM.

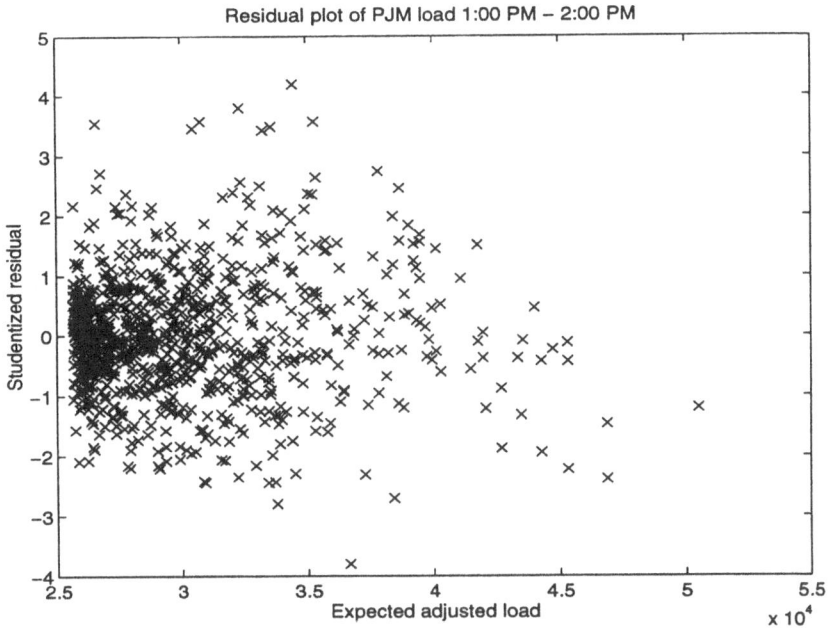

Figure B.5: Residual plot for 1:00 PM - 2:00 PM.

Studentized Residual	Day		Date	Adjusted Load		High	Low
				Actual	Expected		
3.546	28	Fri	1/28/94	31286	26609	58	33
3.423	213	Mon	8/1/94	37783	33285	80	71
3.456	217	Fri	8/5/94	35046	30488	86	60
3.566	226	Sun	8/14/94	39988	35282	90	69
−3.798	622	Thu	9/14/95	31640	36642	92	70
4.190	737	Sun	1/7/96	39938	34473	22	12
3.792	872	Tue	5/21/96	37333	32371	91	60
3.484	899	Mon	6/17/96	38206	33604	88	67
3.572	906	Mon	6/24/96	35521	30803	84	63

Table B.14: List of cases with high residuals, 1:00 PM - 2:00 PM.

Results of Regression of Load vs. Temperature

r_s	Day		Date	Prev. Day				Next Day	
				Low	High	Low	High	Low	High
3.546	28	Fri	1/28/94	11	33	33	58	34	42
3.423	213	Mon	8/1/94	72	84	71	80	74	88
3.456	217	Fri	8/5/94	76	91	60	86	55	77
3.566	226	Sun	8/14/94	75	92	69	90	62	75
−3.798	622	Thu	9/14/95	69	88	70	92	61	80
4.190	737	Sun	1/7/96	10	20	12	22	14	26
3.792	872	Tue	5/21/96	68	94	60	91	55	80
3.484	899	Mon	6/17/96	66	91	67	88	70	82
3.572	906	Mon	6/24/96	68	81	63	84	68	87

Table B.15: List of cases with high residuals, 1:00 PM - 2:00 PM.

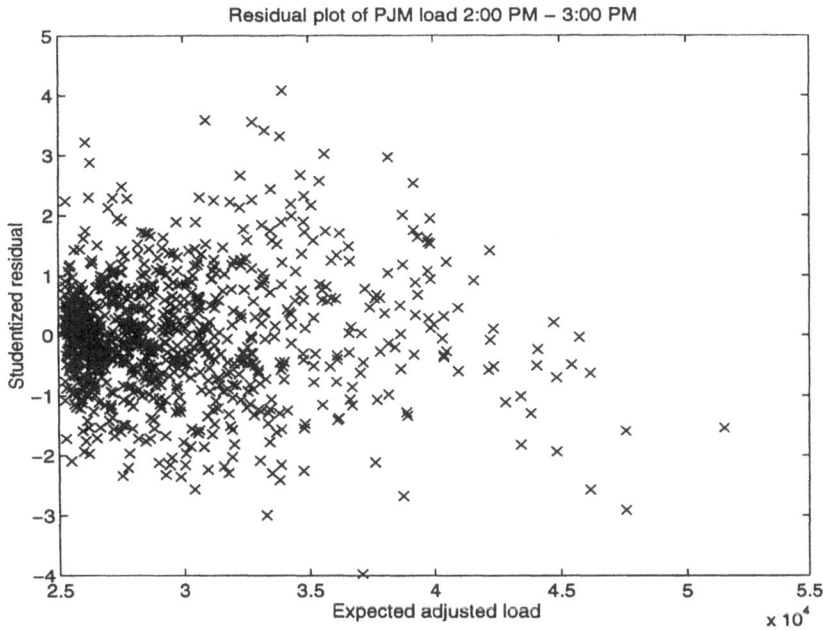

Figure B.6: Residual plot for 2:00 PM - 3:00 PM.

Studentized Residual	Day		Date	Adjusted Load Actual	Adjusted Load Expected	High	Low
3.219	28	Fri	1/28/94	30504	26079	58	33
3.417	213	Mon	8/1/94	37915	33241	80	71
3.022	226	Sun	8/14/94	39781	35621	90	69
−3.974	622	Thu	9/14/95	31630	37075	92	70
4.079	737	Sun	1/7/96	39491	33951	22	12
3.558	872	Tue	5/21/96	37615	32765	91	60
3.318	899	Mon	6/17/96	38425	33860	88	67
3.588	906	Mon	6/24/96	35841	30907	84	63

Table B.16: List of cases with high residuals, 2:00 PM - 3:00 PM.

r_s	Day		Date	Prev. Day Low	High	Low	High	Next Day Low	High
3.219	28	Fri	1/28/94	11	33	33	58	34	42
3.417	213	Mon	8/1/94	72	84	71	80	74	88
3.022	226	Sun	8/14/94	75	92	69	90	62	75
−3.974	622	Thu	9/14/95	69	88	70	92	61	80
4.079	737	Sun	1/7/96	10	20	12	22	14	26
3.558	872	Tue	5/21/96	68	94	60	91	55	80
3.318	899	Mon	6/17/96	66	91	67	88	70	82
3.588	906	Mon	6/24/96	68	81	63	84	68	87

Table B.17: List of cases with high residuals, 2:00 PM - 3:00 PM.

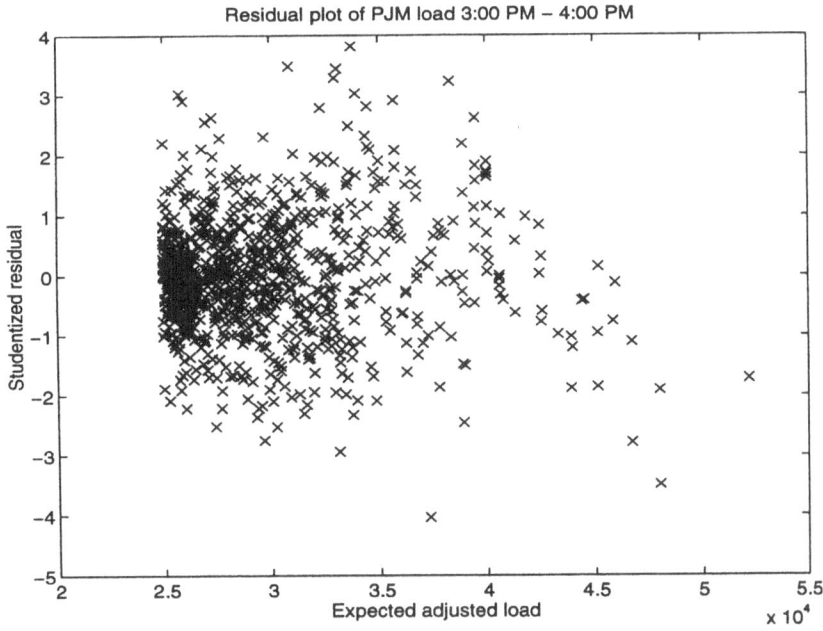

Figure B.7: Residual plot for 3:00 PM - 4:00 PM.

Studentized Residual	Day		Date	Adjusted Load Actual	Adjusted Load Expected	High	Low
3.015	28	Fri	1/28/94	30078	25774	58	33
−3.489	189	Fri	7/8/94	43119	48040	99	80
3.441	213	Mon	8/1/94	37983	33101	80	71
3.230	591	Mon	8/14/95	42924	38318	91	73
−4.036	622	Thu	9/14/95	31560	37294	92	70
3.811	737	Sun	1/7/96	39153	33776	22	12
3.287	872	Tue	5/21/96	37640	32987	91	60
3.027	899	Mon	6/17/96	38278	33952	88	67
3.479	906	Mon	6/24/96	35866	30901	84	63

Table B.18: List of cases with high residuals, 3:00 PM - 4:00 PM.

r_s	Day		Date	Prev. Day				Next Day	
				Low	High	Low	High	Low	High
3.015	28	Fri	1/28/94	11	33	33	58	34	42
−3.489	189	Fri	7/8/94	74	98	80	99	80	99
3.441	213	Mon	8/1/94	72	84	71	80	74	88
3.230	591	Mon	8/14/95	74	91	73	91	74	92
−4.036	622	Thu	9/14/95	69	88	70	92	61	80
3.811	737	Sun	1/7/96	10	20	12	22	14	26
3.287	872	Tue	5/21/96	68	94	60	91	55	80
3.027	899	Mon	6/17/96	66	91	67	88	70	82
3.479	906	Mon	6/24/96	68	81	63	84	68	87

Table B.19: List of cases with high residuals, 3:00 PM - 4:00 PM.

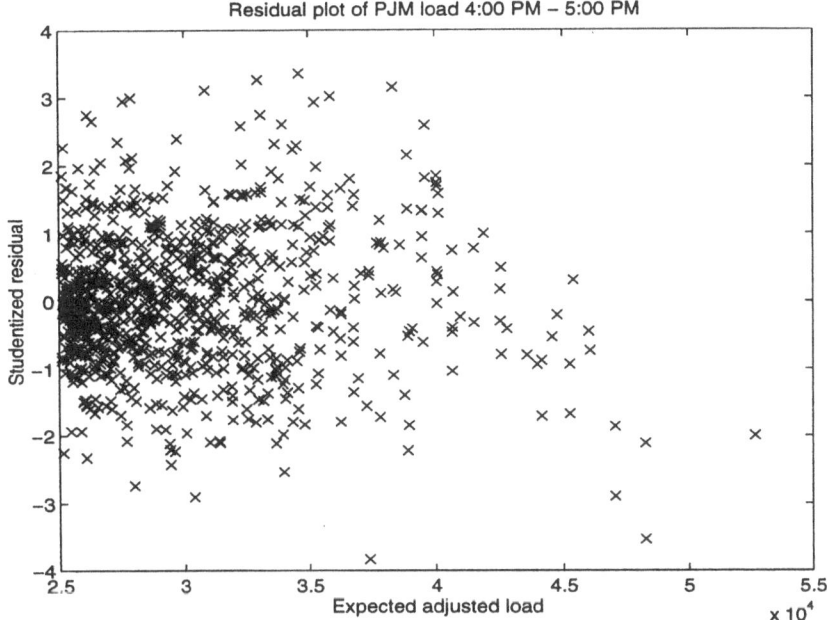

Figure B.8: Residual plot for 4:00 PM - 5:00 PM.

Studentized Residual	Day	Date		Adjusted Load Actual	Expected	High	Low
−3.533	189	Fri	7/8/94	42949	48307	99	80
3.268	213	Mon	8/1/94	37974	32983	80	71
3.004	331	Sun	11/27/94	32492	27873	49	31
3.157	591	Mon	8/14/95	43182	38338	91	73
3.021	598	Mon	8/21/95	40465	35859	93	65
−3.827	622	Thu	9/14/95	31512	37367	92	70
3.361	737	Sun	1/7/96	39739	34630	22	12
3.113	906	Mon	6/24/96	35651	30867	84	63

Table B.20: List of cases with high residuals, 4:00 PM - 5:00 PM.

r_s	Day		Date	Prev. Day				Next Day	
				Low	High	Low	High	Low	High
−3.533	189	Fri	7/8/94	74	98	80	99	80	99
3.268	213	Mon	8/1/94	72	84	71	80	74	88
3.004	331	Sun	11/27/94	32	49	31	49	49	67
3.157	591	Mon	8/14/95	74	91	73	91	74	92
3.021	598	Mon	8/21/95	64	88	65	93	72	86
−3.827	622	Thu	9/14/95	69	88	70	92	61	80
3.361	737	Sun	1/7/96	10	20	12	22	14	26
3.113	906	Mon	6/24/96	68	81	63	84	68	87

Table B.21: List of cases with high residuals, 4:00 PM - 5:00 PM.

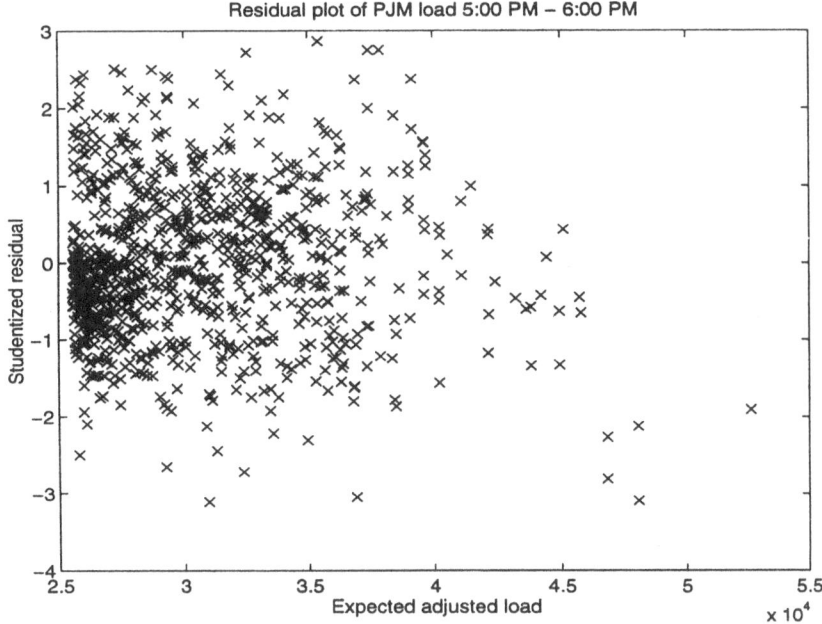

Figure B.9: Residual plot for 5:00 PM - 6:00 PM.

Studentized Residual	Day	Date		Adjusted Load Actual	Expected	High	Low
−3.091	189	Fri	7/8/94	42531	48084	99	80
−3.046	622	Thu	9/14/95	31361	36887	92	70
−3.111	828	Sun	4/7/96	25331	30968	41	36

Table B.22: List of cases with high residuals, 5:00 PM - 6:00 PM.

r_s	Day	Date		Prev. Day Low	High	Low	High	Next Day Low	High
−3.091	189	Fri	7/8/94	74	98	80	99	80	99
−3.046	622	Thu	9/14/95	69	88	70	92	61	80
−3.111	828	Sun	4/7/96	36	50	36	41	33	48

Table B.23: List of cases with high residuals, 5:00 PM - 6:00 PM.

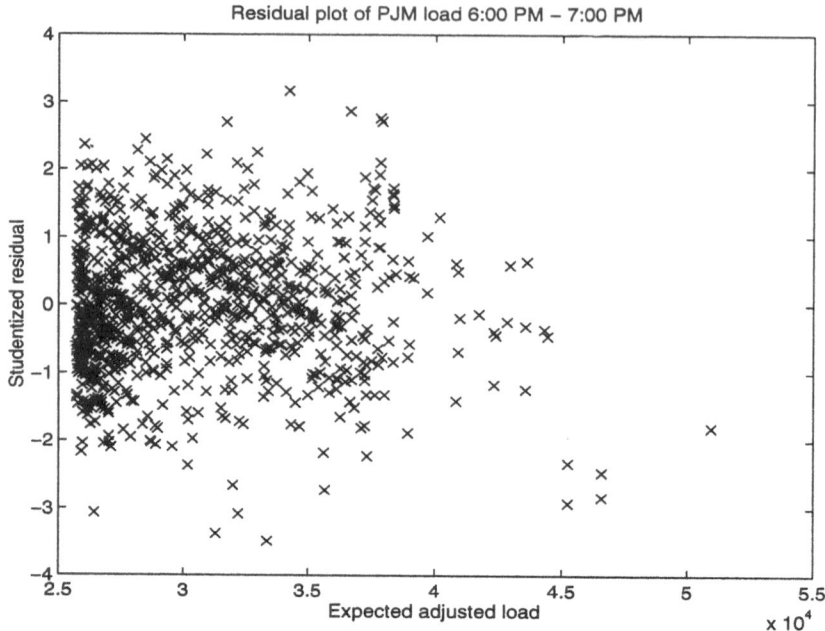

Figure B.10: Residual plot for 6:00 PM - 7:00 PM.

Studentized Residual	Day		Date	Adjusted Load Actual	Expected	High	Low
−3.494	328	Thu	11/24/94	27324	33380	40	24
−3.086	460	Wed	4/5/95	26874	32232	43	27
−3.068	471	Sun	4/16/95	21083	26397	67	38
3.168	598	Mon	8/21/95	39684	34227	93	65
−3.381	828	Sun	4/7/96	25461	31311	41	36

Table B.24: List of cases with high residuals, 6:00 PM - 7:00 PM.

r_s	Day		Date	Prev. Day Low	High	Low	High	Next Day Low	High
−3.494	328	Thu	11/24/94	30	47	24	40	37	54
−3.086	460	Wed	4/5/95	33	69	27	43	33	54
−3.068	471	Sun	4/16/95	40	60	38	67	43	64
3.168	598	Mon	8/21/95	64	88	65	93	72	86
−3.381	828	Sun	4/7/96	36	50	36	41	33	48

Table B.25: List of cases with high residuals, 6:00 PM - 7:00 PM.

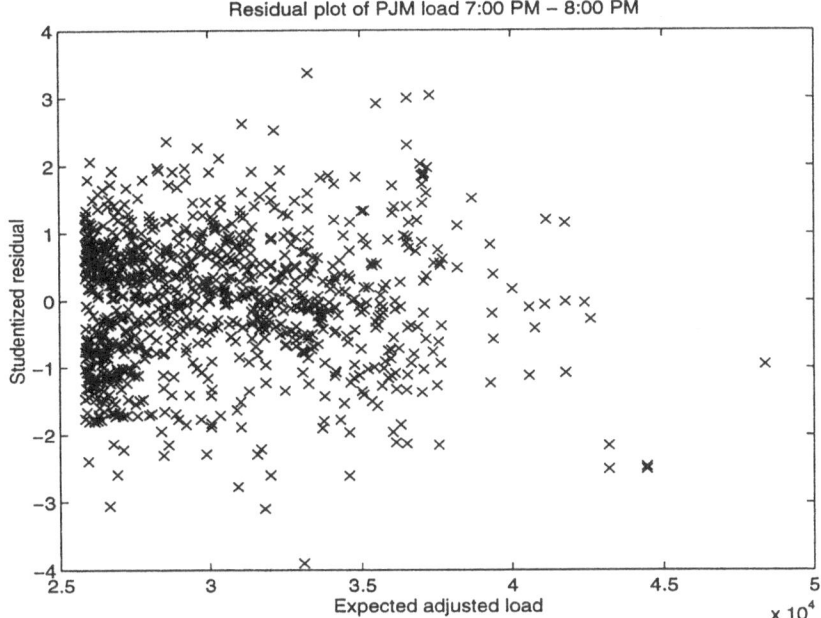

Figure B.11: Residual plot for 7:00 PM - 8:00 PM.

Studentized Residual	Day	Date		Adjusted Load Actual	Expected	High	Low
3.037	15	Sat	1/15/94	41944	37308	17	6
−3.903	328	Thu	11/24/94	27021	33088	40	24
−3.061	471	Sun	4/16/95	21893	26656	67	38
3.370	598	Mon	8/21/95	38499	33290	93	65
−3.100	692	Thu	11/23/95	26989	31823	44	27

Table B.26: List of cases with high residuals, 7:00 PM - 8:00 PM.

r_s	Day	Date		Prev. Day Low	High	Low	High	Next Day Low	High
3.037	15	Sat	1/15/94	17	39	6	17	4	15
−3.903	328	Thu	11/24/94	30	47	24	40	37	54
−3.061	471	Sun	4/16/95	40	60	38	67	43	64
3.370	598	Mon	8/21/95	64	88	65	93	72	86
−3.100	692	Thu	11/23/95	30	45	27	44	34	42

Table B.27: List of cases with high residuals, 7:00 PM - 8:00 PM.

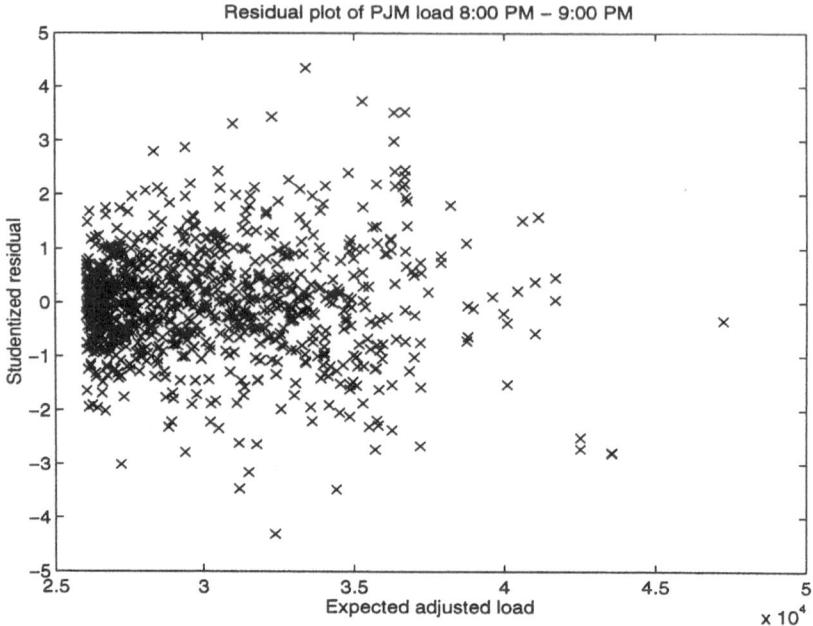

Figure B.12: Residual plot for 8:00 PM - 9:00 PM.

Studentized Residual	Day	Date		Adjusted Load Actual	Expected	High	Low
3.528	15	Sat	1/15/94	41112	36686	17	6
−3.013	149	Sun	5/29/94	23363	27230	80	55
−3.477	184	Sun	7/3/94	29946	34400	90	72
3.307	213	Mon	8/1/94	35216	30997	80	71
−4.302	328	Thu	11/24/94	26883	32378	40	24
−3.160	548	Sun	7/2/95	27471	31521	84	70
3.513	578	Tue	8/1/95	40809	36319	94	72
3.727	591	Mon	8/14/95	40040	35272	91	73
4.341	598	Mon	8/21/95	38887	33386	93	65
3.436	608	Thu	8/31/95	36666	32270	90	66
−3.462	692	Thu	11/23/95	26774	31212	44	27

Table B.28: List of cases with high residuals, 8:00 PM - 9:00 PM.

r_s	Day		Date	Prev. Day Low	High	Low	High	Next Day Low	High
3.528	15	Sat	1/15/94	17	39	6	17	4	15
−3.013	149	Sun	5/29/94	48	71	55	80	59	85
−3.477	184	Sun	7/3/94	71	94	72	90	70	86
3.307	213	Mon	8/1/94	72	84	71	80	74	88
−4.302	328	Thu	11/24/94	30	47	24	40	37	54
−3.160	548	Sun	7/2/95	71	89	70	84	64	83
3.513	578	Tue	8/1/95	73	96	72	94	78	98
3.727	591	Mon	8/14/95	74	91	73	91	74	92
4.341	598	Mon	8/21/95	64	88	65	93	72	86
3.436	608	Thu	8/31/95	68	90	66	90	75	90
−3.462	692	Thu	11/23/95	30	45	27	44	34	42

Table B.29: List of cases with high residuals, 8:00 PM - 9:00 PM.

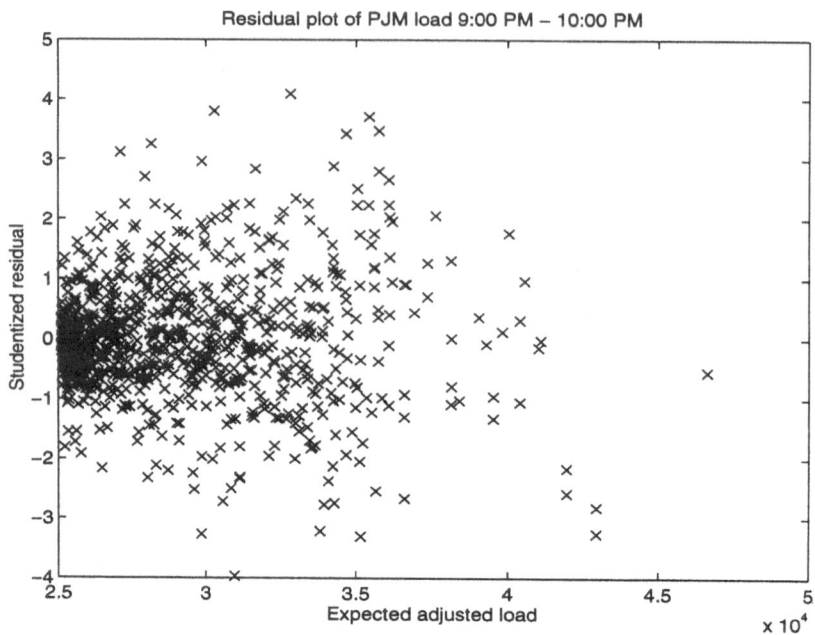

Figure B.13: Residual plot for 9:00 PM - 10:00 PM.

Studentized Residual	Day		Date	Actual	Expected	High	Low
				\multicolumn{2}{}{Adjusted Load}		
3.709	15	Sat	1/15/94	39823	35429	17	6
−3.220	184	Sun	7/3/94	29896	33797	90	72
−3.256	190	Sat	7/9/94	39050	42953	99	80
3.801	213	Mon	8/1/94	34868	30296	80	71
−3.970	328	Thu	11/24/94	26186	30985	40	24
3.470	578	Tue	8/1/95	39945	35754	94	72
3.421	591	Mon	8/14/95	38813	34673	91	73
4.088	598	Mon	8/21/95	37719	32819	93	65
−3.304	609	Fri	9/1/95	31162	35152	90	75
−3.268	692	Thu	11/23/95	25889	29850	44	27
3.250	749	Fri	1/19/96	32040	28158	62	22
3.115	757	Sat	1/27/96	30887	27114	57	32

Table B.30: List of cases with high residuals, 9:00 PM - 10:00 PM.

r_s	Day		Date	Prev. Day Low	Prev. Day High	Low	High	Next Day Low	Next Day High
3.709	15	Sat	1/15/94	17	39	6	17	4	15
−3.220	184	Sun	7/3/94	71	94	72	90	70	86
−3.256	190	Sat	7/9/94	80	99	80	99	76	92
3.801	213	Mon	8/1/94	72	84	71	80	74	88
−3.970	328	Thu	11/24/94	30	47	24	40	37	54
3.470	578	Tue	8/1/95	73	96	72	94	78	98
3.421	591	Mon	8/14/95	74	91	73	91	74	92
4.088	598	Mon	8/21/95	64	88	65	93	72	86
−3.304	609	Fri	9/1/95	66	90	75	90	69	83
−3.268	692	Thu	11/23/95	30	45	27	44	34	42
3.250	749	Fri	1/19/96	29	58	22	62	19	29
3.115	757	Sat	1/27/96	22	48	32	57	24	36

Table B.31: List of cases with high residuals, 9:00 PM - 10:00 PM.

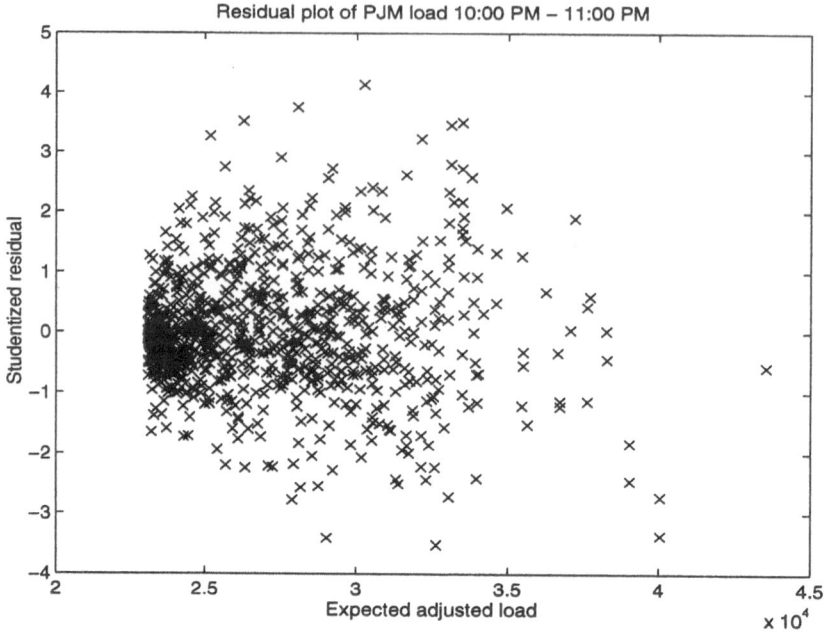

Figure B.14: Residual plot for 10:00 PM - 11:00 PM.

Studentized Residual	Day		Date	Adjusted Load Actual	Expected	High	Low
3.507	15	Sat	1/15/94	37532	33498	17	6
−3.366	190	Sat	7/9/94	36131	40043	99	80
3.752	213	Mon	8/1/94	32456	28076	80	71
−3.409	328	Thu	11/24/94	25023	29029	40	24
3.462	578	Tue	8/1/95	37175	33118	94	72
3.232	591	Mon	8/14/95	35937	32140	91	73
4.134	598	Mon	8/21/95	35056	30250	93	65
−3.521	609	Fri	9/1/95	28517	32638	90	75
3.522	749	Fri	1/19/96	30373	26295	62	22
3.273	757	Sat	1/27/96	29032	25187	57	32

Table B.32: List of cases with high residuals, 10:00 PM - 11:00 PM.

r_s	Day		Date	Prev. Day				Next Day	
				Low	High	Low	High	Low	High
3.507	15	Sat	1/15/94	17	39	6	17	4	15
−3.366	190	Sat	7/9/94	80	99	80	99	76	92
3.752	213	Mon	8/1/94	72	84	71	80	74	88
−3.409	328	Thu	11/24/94	30	47	24	40	37	54
3.462	578	Tue	8/1/95	73	96	72	94	78	98
3.232	591	Mon	8/14/95	74	91	73	91	74	92
4.134	598	Mon	8/21/95	64	88	65	93	72	86
−3.521	609	Fri	9/1/95	66	90	75	90	69	83
3.522	749	Fri	1/19/96	29	58	22	62	19	29
3.273	757	Sat	1/27/96	22	48	32	57	24	36

Table B.33: List of cases with high residuals, 10:00 PM - 11:00 PM.

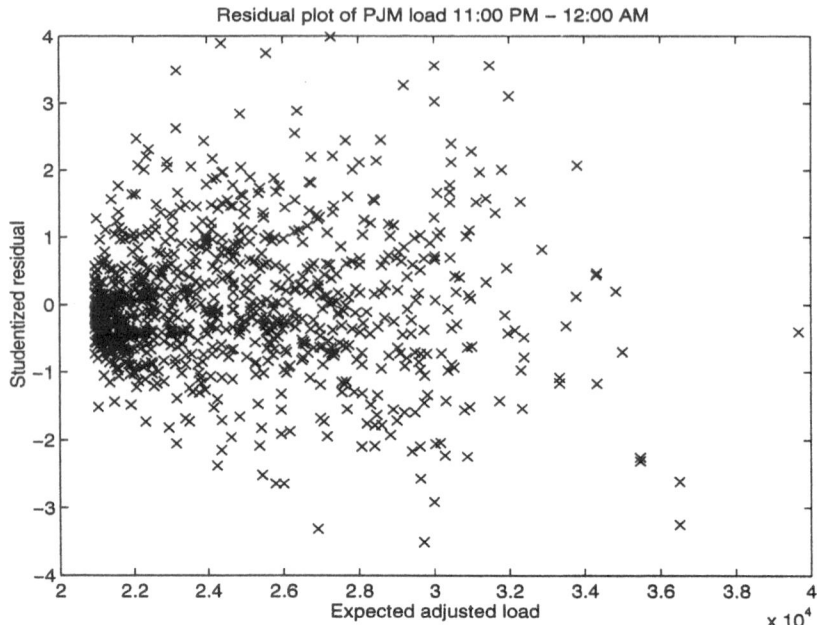

Figure B.15: Residual plot for 11:00 PM - 12:00 AM.

Studentized Residual	Day		Date	Adjusted Load Actual	Expected	High	Low
3.557	15	Sat	1/15/94	35398	31483	17	6
3.104	18	Tue	1/18/94	35073	31992	33	2
−3.249	190	Sat	7/9/94	32894	36510	99	80
3.748	213	Mon	8/1/94	29765	25576	80	71
−3.312	328	Thu	11/24/94	23186	26914	40	24
3.558	578	Tue	8/1/95	34011	30021	94	72
3.267	591	Mon	8/14/95	32869	29195	91	73
3.029	593	Wed	8/16/95	33424	30021	94	72
3.991	598	Mon	8/21/95	31724	27279	93	65
−3.505	609	Fri	9/1/95	25791	29718	90	75
3.892	749	Fri	1/19/96	28695	24388	62	22
3.484	757	Sat	1/27/96	27079	23165	57	32

Table B.34: List of cases with high residuals, 11:00 PM - 12:00 AM.

r_s	Day		Date	Prev. Day Low	High	Low	High	Next Day Low	High
3.557	15	Sat	1/15/94	17	39	6	17	4	15
3.104	18	Tue	1/18/94	14	34	2	33	−5	6
−3.249	190	Sat	7/9/94	80	99	80	99	76	92
3.748	213	Mon	8/1/94	72	84	71	80	74	88
−3.312	328	Thu	11/24/94	30	47	24	40	37	54
3.558	578	Tue	8/1/95	73	96	72	94	78	98
3.267	591	Mon	8/14/95	74	91	73	91	74	92
3.029	593	Wed	8/16/95	74	92	72	94	79	96
3.991	598	Mon	8/21/95	64	88	65	93	72	86
−3.505	609	Fri	9/1/95	66	90	75	90	69	83
3.892	749	Fri	1/19/96	29	58	22	62	19	29
3.484	757	Sat	1/27/96	22	48	32	57	24	36

Table B.35: List of cases with high residuals, 11:00 PM - 12:00 AM.

Hour	Coefficient of T_{lo}^3			Coefficient of T_{lo}^2		
	Value	Lower	Upper	Value	Lower	Upper
12 AM - 1 AM	0.0742	0.0622	0.0861	−4.87	−6.42	−3.31
1 AM - 2 AM	0.0721	0.0607	0.0836	−4.62	−6.11	−3.13
2 AM - 3 AM	0.0717	0.0606	0.0828	−4.56	−6.01	−3.12
3 AM - 4 AM	0.0722	0.0613	0.0831	−4.60	−6.02	−3.18
4 AM - 5 AM	0.0749	0.0641	0.0858	−4.84	−6.25	−3.44
5 AM - 6 AM	0.0795	0.0681	0.0908	−5.39	−6.87	−3.91
6 AM - 7 AM	0.0825	0.0692	0.0958	−5.84	−7.58	−4.11
7 AM - 8 AM	0.0891	0.0749	0.1032	−6.49	−8.32	−4.65
8 AM - 9 AM	0.0923	0.0787	0.1059	−6.42	−8.19	−4.64
9 AM - 10 AM	0.0955	0.0817	0.1093	−6.28	−8.08	−4.49
10 AM - 11 AM	0.1005	0.0857	0.1153	−6.36	−8.29	−4.43
11 AM - 12 PM	0.1045	0.0883	0.1207	−6.33	−8.44	−4.23

Table B.36: Regression results for T_{lo}^3 and T_{lo}^2.

B.2 Regression using Previous Day's High Temperature

As mentioned earlier, superior regression results were obtained for the morning hours by using the high temperature from the previous day (here designated $T_{hi,-1}$) in combination with the same day's low instead of using the high and low temperature for the same day. The resulting regression coefficients and 95% confidence intervals are given in the following tables. A table showing the coefficient of determination, F-statistic, and standard error of regression for each hour's regression is also given. As before, a residual plot and set of possible outliers is presented for each hour.

Hour	Coefficient of T_{lo}			Constant Term		
	Value	Lower	Upper	Value	Lower	Upper
12 AM - 1 AM	−70.53	−132.14	−8.93	28942	27634	30250
1 AM - 2 AM	−88.46	−147.53	−29.38	28592	27338	29847
2 AM - 3 AM	−101.35	−158.55	−44.15	28560	27346	29775
3 AM - 4 AM	−110.39	−166.63	−54.16	28822	27628	30016
4 AM - 5 AM	−116.39	−172.21	−60.56	29169	27983	30354
5 AM - 6 AM	−108.68	−167.34	−50.02	29940	28694	31186
6 AM - 7 AM	−104.77	−173.46	−36.07	31370	29911	32829
7 AM - 8 AM	−82.93	−155.70	−10.16	32533	30988	34079
8 AM - 9 AM	−86.65	−156.96	−16.34	33586	32093	35079
9 AM - 10 AM	−90.90	−161.99	−19.81	34165	32656	35675
10 AM - 11 AM	−90.35	−166.82	−13.87	34298	32674	35922
11 AM - 12 PM	−94.61	−178.01	−11.21	33839	32069	35610

Table B.37: Regression results for T_{lo} and the constant term.

Hour	Coefficient of $T^3_{hi,-1}$			Coefficient of $T^2_{hi,-1}$		
	Value	Lower	Upper	Value	Lower	Upper
12 AM - 1 AM	0.0746	0.0654	0.0837	−9.70	−11.33	−8.06
1 AM - 2 AM	0.0682	0.0595	0.0770	−8.92	−10.49	−7.35
2 AM - 3 AM	0.0634	0.0549	0.0719	−8.34	−9.86	−6.82
3 AM - 4 AM	0.0589	0.0505	0.0673	−7.79	−9.29	−6.30
4 AM - 5 AM	0.0561	0.0478	0.0644	−7.57	−9.06	−6.09
5 AM - 6 AM	0.0558	0.0471	0.0646	−7.84	−9.40	−6.28
6 AM - 7 AM	0.0592	0.0490	0.0694	−8.80	−10.62	−6.97
7 AM - 8 AM	0.0611	0.0503	0.0719	−9.17	−11.10	−7.24
8 AM - 9 AM	0.0607	0.0503	0.0712	−8.93	−10.80	−7.06
9 AM - 10 AM	0.0605	0.0500	0.0711	−8.69	−10.58	−6.81
10 AM - 11 AM	0.0603	0.0490	0.0717	−8.50	−10.53	−6.47
11 AM - 12 PM	0.0615	0.0491	0.0739	−8.60	−10.82	−6.39

Table B.38: Regression results for $T^3_{hi,-1}$ and $T^2_{hi,-1}$.

Hour	Coefficient of $T_{hi,-1}$		
	Value	Lower	Upper
12 AM - 1 AM	265.35	173.05	357.66
1 AM - 2 AM	241.58	153.06	330.10
2 AM - 3 AM	226.96	141.25	312.68
3 AM - 4 AM	211.52	127.26	295.79
4 AM - 5 AM	216.78	133.12	300.43
5 AM - 6 AM	249.48	161.58	337.38
6 AM - 7 AM	323.71	220.78	426.64
7 AM - 8 AM	351.38	242.34	460.42
8 AM - 9 AM	333.87	228.52	439.22
9 AM - 10 AM	313.30	206.78	419.83
10 AM - 11 AM	298.59	184.00	413.18
11 AM - 12 PM	301.60	176.63	426.56

Table B.39: Regression results for $T_{hi,-1}$.

Hour	R^2	F	$\hat{\sigma}^2$	Standard Error
12 AM - 1 AM	0.9105	1527	991996	995.99
1 AM - 2 AM	0.9123	1561	912281	955.13
2 AM - 3 AM	0.9152	1617	855197	924.77
3 AM - 4 AM	0.9173	1663	826594	909.17
4 AM - 5 AM	0.9196	1715	814701	902.61
5 AM - 6 AM	0.9148	1611	899475	948.41
6 AM - 7 AM	0.8932	1254	1233436	1110.60
7 AM - 8 AM	0.8760	1060	1384205	1176.52
8 AM - 9 AM	0.8769	1068	1292103	1136.71
9 AM - 10 AM	0.8784	1083	1321107	1149.39
10 AM - 11 AM	0.8760	1060	1528711	1236.41
11 AM - 12 PM	0.8749	1049	1818077	1348.36

Table B.40: Regression statistics and estimates of variance.

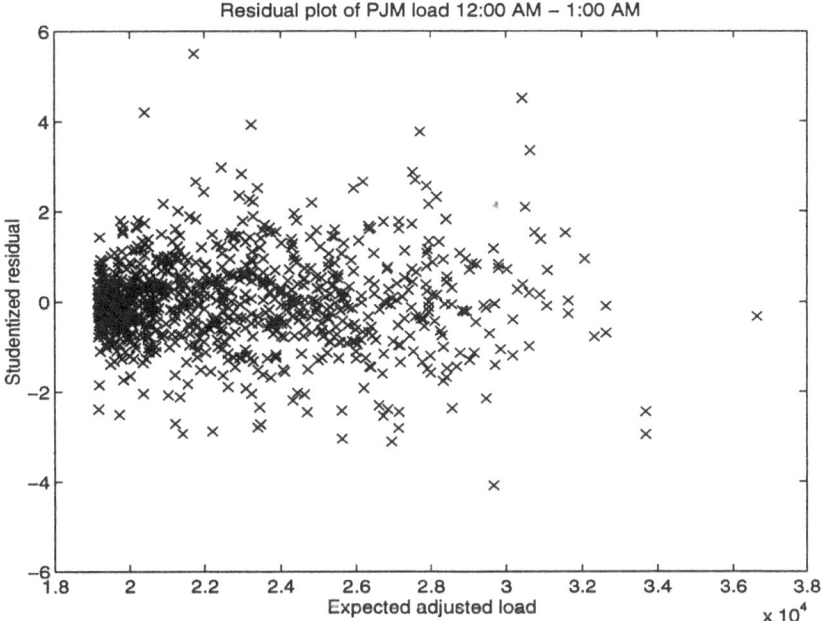

Figure B.16: Residual plot for 12:00 AM - 1:00 AM.

Studentized Residual	Day	Date		Adjusted Load Actual	Expected	High	Low
3.359	16	Sun	1/16/94	33895	30637	17	4
−4.076	18	Tue	1/18/94	25728	29659	34	2
4.522	19	Wed	1/19/94	33749	30423	33	−5
5.493	217	Fri	8/5/94	28580	21733	91	60
4.197	241	Mon	8/29/94	26696	20413	89	60
−3.046	551	Wed	7/5/95	21457	25633	86	71
−3.105	725	Tue	12/26/95	23760	26953	32	22
3.778	750	Sat	1/20/96	26982	27712	62	19
3.929	872	Tue	5/21/96	28222	23240	94	60

Table B.41: List of cases with high residuals, 12:00 AM - 1:00 AM.

| | | | | Prev. Day | | | | Next Day | |
r_s	Day		Date	Low	High	Low	High	Low	High
3.359	16	Sun	1/16/94	6	17	4	15	14	34
−4.076	18	Tue	1/18/94	14	34	2	33	−5	6
4.522	19	Wed	1/19/94	2	33	−5	6	1	15
5.493	217	Fri	8/5/94	76	91	60	86	55	77
4.197	241	Mon	8/29/94	70	89	60	79	59	79
−3.046	551	Wed	7/5/95	65	86	71	90	73	90
−3.105	725	Tue	12/26/95	27	32	22	31	21	31
3.778	750	Sat	1/20/96	22	62	19	29	22	32
3.929	872	Tue	5/21/96	68	94	60	91	55	80

Table B.42: List of cases with high residuals, 12:00 AM - 1:00 AM.

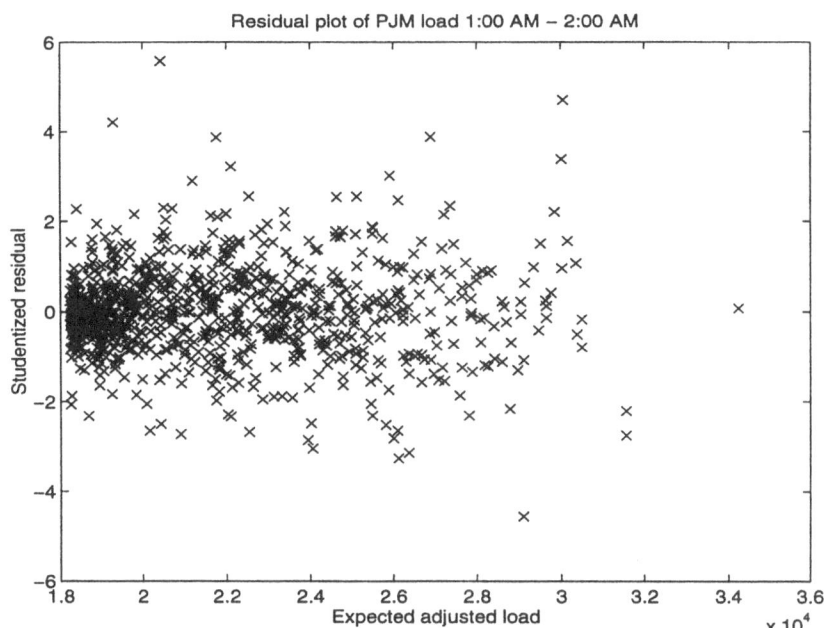

Figure B.17: Residual plot for 1:00 AM - 2:00 AM.

Studentized Residual	Day		Date	Adjusted Load Actual	Expected	High	Low
3.399	16	Sun	1/16/94	33171	30016	17	4
−4.555	18	Tue	1/18/94	24922	29108	34	2
4.720	19	Wed	1/19/94	33329	30053	33	−5
5.582	217	Fri	8/5/94	26965	20430	91	60
4.208	241	Mon	8/29/94	25135	19281	89	60
−3.128	369	Wed	1/4/95	23791	26377	32	18
−3.039	551	Wed	7/5/95	20146	24058	86	71
3.020	562	Sun	7/16/95	33987	25920	103	75
−3.253	725	Tue	12/26/95	22928	26127	32	22
3.889	750	Sat	1/20/96	26324	26906	62	19
3.225	780	Mon	2/19/96	26626	22111	31	37
3.873	872	Tue	5/21/96	26410	21763	94	60

Table B.43: List of cases with high residuals, 1:00 AM - 2:00 AM.

r_s	Day		Date	Prev. Day				Next Day	
				Low	High	Low	High	Low	High
3.399	16	Sun	1/16/94	6	17	4	15	14	34
−4.555	18	Tue	1/18/94	14	34	2	33	−5	6
4.720	19	Wed	1/19/94	2	33	−5	6	1	15
5.582	217	Fri	8/5/94	76	91	60	86	55	77
4.208	241	Mon	8/29/94	70	89	60	79	59	79
−3.128	369	Wed	1/4/95	20	32	18	35	14	25
−3.039	551	Wed	7/5/95	65	86	71	90	73	90
3.020	562	Sun	7/16/95	81	103	75	91	73	94
−3.253	725	Tue	12/26/95	27	32	22	31	21	31
3.889	750	Sat	1/20/96	22	62	19	29	22	32
3.225	780	Mon	2/19/96	16	31	37	42	48	59
3.873	872	Tue	5/21/96	68	94	60	91	55	80

Table B.44: List of cases with high residuals, 1:00 AM - 2:00 AM.

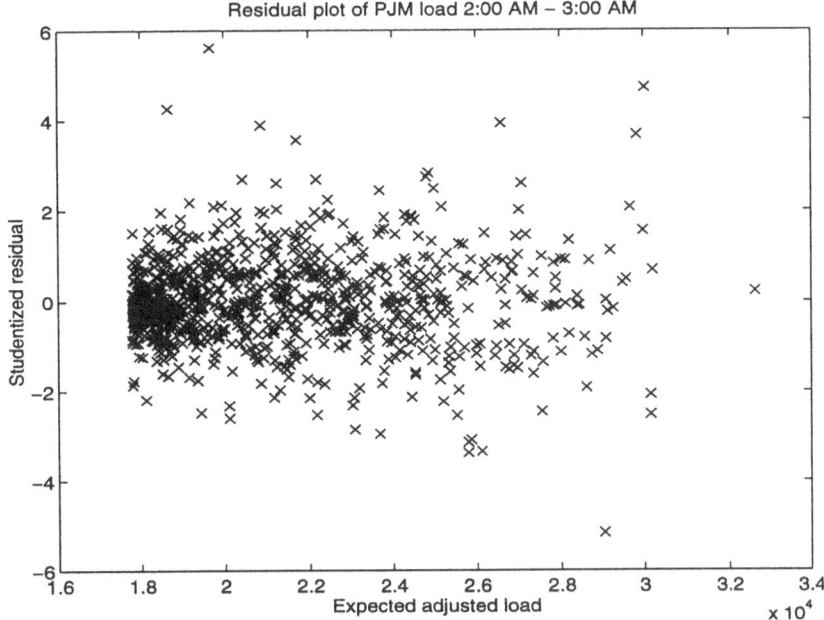

Figure B.18: Residual plot for 2:00 AM - 3:00 AM.

Studentized Residual	Day		Date	Adjusted Load Actual	Adjusted Load Expected	High	Low
−3.079	8	Sat	1/8/94	23534	25864	33	18
−3.132	14	Fri	1/14/94	22656	25794	41	17
3.682	16	Sun	1/16/94	33132	29828	17	4
−5.156	18	Tue	1/18/94	24469	29023	34	2
4.736	19	Wed	1/19/94	33230	30019	33	−5
5.619	217	Fri	8/5/94	25935	19667	91	60
4.258	241	Mon	8/29/94	24215	18650	89	60
−3.339	369	Wed	1/4/95	23412	26118	32	18
−3.373	725	Tue	12/26/95	22591	25793	32	22
3.942	750	Sat	1/20/96	26153	26592	62	19
3.564	780	Mon	2/19/96	26344	21708	31	37
3.891	872	Tue	5/21/96	25317	20867	94	60

Table B.45: List of cases with high residuals, 2:00 AM - 3:00 AM.

r_s	Day		Date	Prev. Day				Next Day	
				Low	High	Low	High	Low	High
−3.079	8	Sat	1/8/94	30	33	18	37	17	28
−3.132	14	Fri	1/14/94	36	41	17	39	6	17
3.682	16	Sun	1/16/94	6	17	4	15	14	34
−5.156	18	Tue	1/18/94	14	34	2	33	−5	6
4.736	19	Wed	1/19/94	2	33	−5	6	1	15
5.619	217	Fri	8/5/94	76	91	60	86	55	77
4.258	241	Mon	8/29/94	70	89	60	79	59	79
−3.339	369	Wed	1/4/95	20	32	18	35	14	25
−3.373	725	Tue	12/26/95	27	32	22	31	21	31
3.942	750	Sat	1/20/96	22	62	19	29	22	32
3.564	780	Mon	2/19/96	16	31	37	42	48	59
3.891	872	Tue	5/21/96	68	94	60	91	55	80

Table B.46: List of cases with high residuals, 2:00 AM - 3:00 AM.

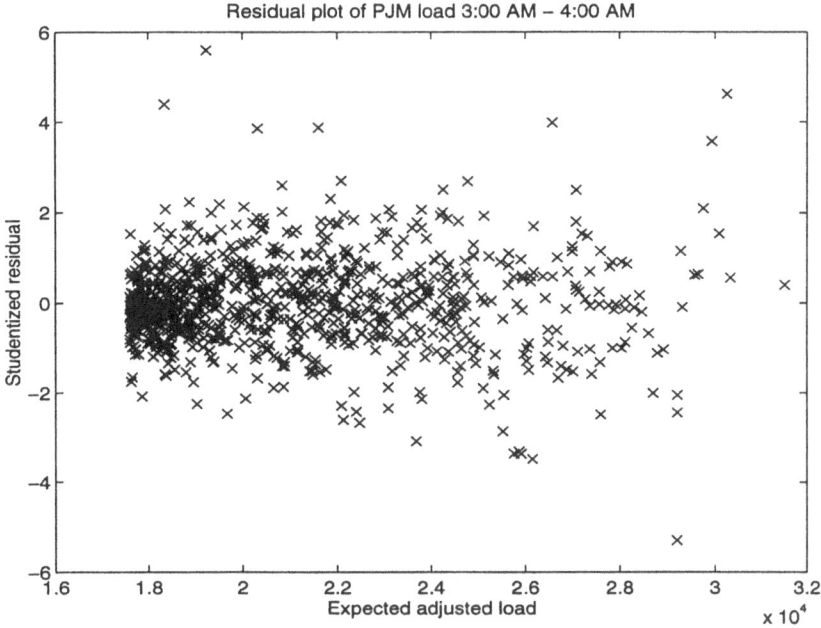

Figure B.19: Residual plot for 3:00 AM - 4:00 AM.

Studentized Residual	Day	Date		Adjusted Load Actual	Expected	High	Low
−3.359	8	Sat	1/8/94	23347	25905	33	18
−3.311	14	Fri	1/14/94	22624	25859	41	17
3.578	16	Sun	1/16/94	33086	29930	17	4
−5.288	18	Tue	1/18/94	24611	29193	34	2
4.631	19	Wed	1/19/94	33324	30252	33	−5
5.604	217	Fri	8/5/94	25288	19237	91	60
4.395	241	Mon	8/29/94	23773	18336	89	60
−3.081	367	Mon	1/2/95	19626	23682	53	25
−3.480	369	Wed	1/4/95	23348	26146	32	18
−3.358	725	Tue	12/26/95	22626	25758	32	22
3.989	750	Sat	1/20/96	26294	26575	62	19
3.876	780	Mon	2/19/96	26379	21595	31	37
3.861	872	Tue	5/21/96	24595	20318	94	60

Table B.47: List of cases with high residuals, 3:00 AM - 4:00 AM.

r_s	Day		Date	Prev. Day Low	High	Low	High	Next Day Low	High
−3.359	8	Sat	1/8/94	30	33	18	37	17	28
−3.311	14	Fri	1/14/94	36	41	17	39	6	17
3.578	16	Sun	1/16/94	6	17	4	15	14	34
−5.288	18	Tue	1/18/94	14	34	2	33	−5	6
4.631	19	Wed	1/19/94	2	33	−5	6	1	15
5.604	217	Fri	8/5/94	76	91	60	86	55	77
4.395	241	Mon	8/29/94	70	89	60	79	59	79
−3.081	367	Mon	1/2/95	37	53	25	43	20	32
−3.480	369	Wed	1/4/95	20	32	18	35	14	25
−3.358	725	Tue	12/26/95	27	32	22	31	21	31
3.989	750	Sat	1/20/96	22	62	19	29	22	32
3.876	780	Mon	2/19/96	16	31	37	42	48	59
3.861	872	Tue	5/21/96	68	94	60	91	55	80

Table B.48: List of cases with high residuals, 3:00 AM - 4:00 AM.

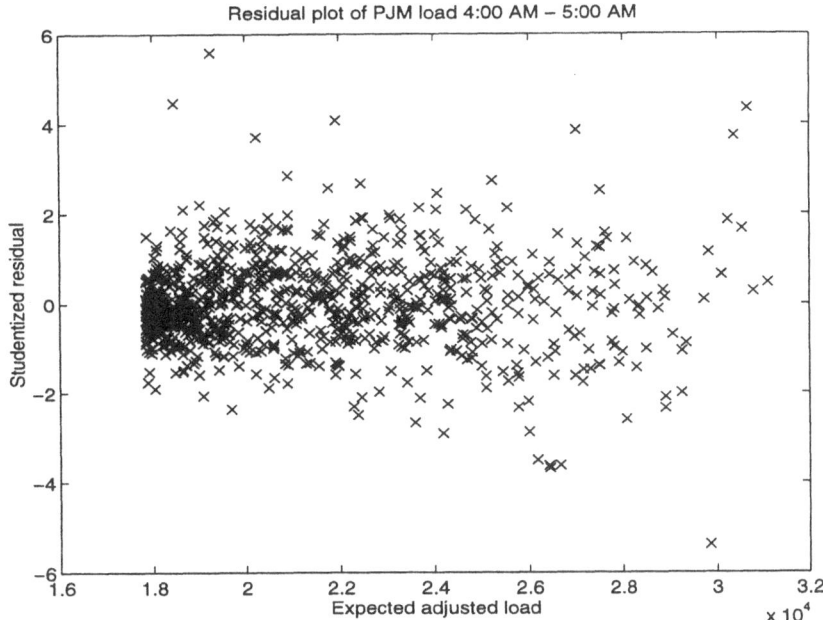

Figure B.20: Residual plot for 4:00 AM - 5:00 AM.

Studentized Residual	Day		Date	Adjusted Load Actual	Adjusted Load Expected	High	Low
−3.685	8	Sat	1/8/94	23570	26436	33	18
−3.627	14	Fri	1/14/94	22949	26422	41	17
3.746	16	Sun	1/16/94	33666	30367	17	4
−5.381	18	Tue	1/18/94	25226	29840	34	2
4.358	19	Wed	1/19/94	33784	30660	33	−5
5.584	217	Fri	8/5/94	25157	19262	91	60
4.468	241	Mon	8/29/94	23774	18464	89	60
−3.621	369	Wed	1/4/95	23724	26657	32	18
−3.507	725	Tue	12/26/95	22945	26175	32	22
3.869	750	Sat	1/20/96	26843	27008	62	19
4.078	780	Mon	2/19/96	26751	21929	31	37
3.713	872	Tue	5/21/96	24289	20240	94	60

Table B.49: List of cases with high residuals, 4:00 AM - 5:00 AM.

r_s	Day		Date	Prev. Day Low	Prev. Day High	Low	High	Next Day Low	Next Day High
−3.685	8	Sat	1/8/94	30	33	18	37	17	28
−3.627	14	Fri	1/14/94	36	41	17	39	6	17
3.746	16	Sun	1/16/94	6	17	4	15	14	34
−5.381	18	Tue	1/18/94	14	34	2	33	−5	6
4.358	19	Wed	1/19/94	2	33	−5	6	1	15
5.584	217	Fri	8/5/94	76	91	60	86	55	77
4.468	241	Mon	8/29/94	70	89	60	79	59	79
−3.621	369	Wed	1/4/95	20	32	18	35	14	25
−3.507	725	Tue	12/26/95	27	32	22	31	21	31
3.869	750	Sat	1/20/96	22	62	19	29	22	32
4.078	780	Mon	2/19/96	16	31	37	42	48	59
3.713	872	Tue	5/21/96	68	94	60	91	55	80

Table B.50: List of cases with high residuals, 4:00 AM - 5:00 AM.

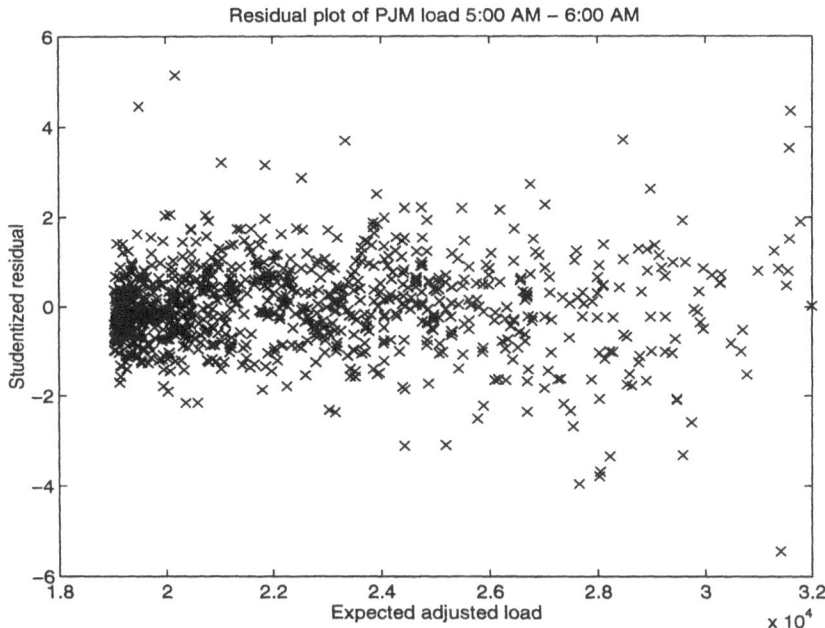

Figure B.21: Residual plot for 5:00 AM - 6:00 AM.

Studentized Residual	Day		Date	Adjusted Load Actual	Expected	High	Low
−3.781	8	Sat	1/8/94	24855	28030	33	18
−3.678	14	Fri	1/14/94	24383	28046	41	17
4.355	16	Sun	1/16/94	35650	31591	17	4
−5.438	18	Tue	1/18/94	26529	31407	34	2
3.537	19	Wed	1/19/94	34818	31565	33	−5
5.151	217	Fri	8/5/94	25828	20176	91	60
4.455	241	Mon	8/29/94	24816	19495	89	60
−3.095	363	Thu	12/29/94	21160	25193	56	26
−3.344	369	Wed	1/4/95	25343	28225	32	18
3.157	583	Sun	8/6/95	27022	21866	94	71
−3.953	725	Tue	12/26/95	23864	27651	32	22
−3.318	739	Tue	1/9/96	26638	29585	26	14
−3.113	749	Fri	1/19/96	21874	24435	58	22
3.723	750	Sat	1/20/96	28578	28479	62	19
3.697	780	Mon	2/19/96	27872	23351	31	37
3.212	872	Tue	5/21/96	24736	21050	94	60

Table B.51: List of cases with high residuals, 5:00 AM - 6:00 AM.

r_s	Day		Date	Prev. Day				Next Day	
				Low	High	Low	High	Low	High
−3.781	8	Sat	1/8/94	30	33	18	37	17	28
−3.678	14	Fri	1/14/94	36	41	17	39	6	17
4.355	16	Sun	1/16/94	6	17	4	15	14	34
−5.438	18	Tue	1/18/94	14	34	2	33	−5	6
3.537	19	Wed	1/19/94	2	33	−5	6	1	15
5.151	217	Fri	8/5/94	76	91	60	86	55	77
4.455	241	Mon	8/29/94	70	89	60	79	59	79
−3.095	363	Thu	12/29/94	31	56	26	46	22	38
−3.344	369	Wed	1/4/95	20	32	18	35	14	25
3.157	583	Sun	8/6/95	77	94	71	80	68	81
−3.953	725	Tue	12/26/95	27	32	22	31	21	31
−3.318	739	Tue	1/9/96	14	26	14	29	20	32
−3.113	749	Fri	1/19/96	29	58	22	62	19	29
3.723	750	Sat	1/20/96	22	62	19	29	22	32
3.697	780	Mon	2/19/96	16	31	37	42	48	59
3.212	872	Tue	5/21/96	68	94	60	91	55	80

Table B.52: List of cases with high residuals, 5:00 AM - 6:00 AM.

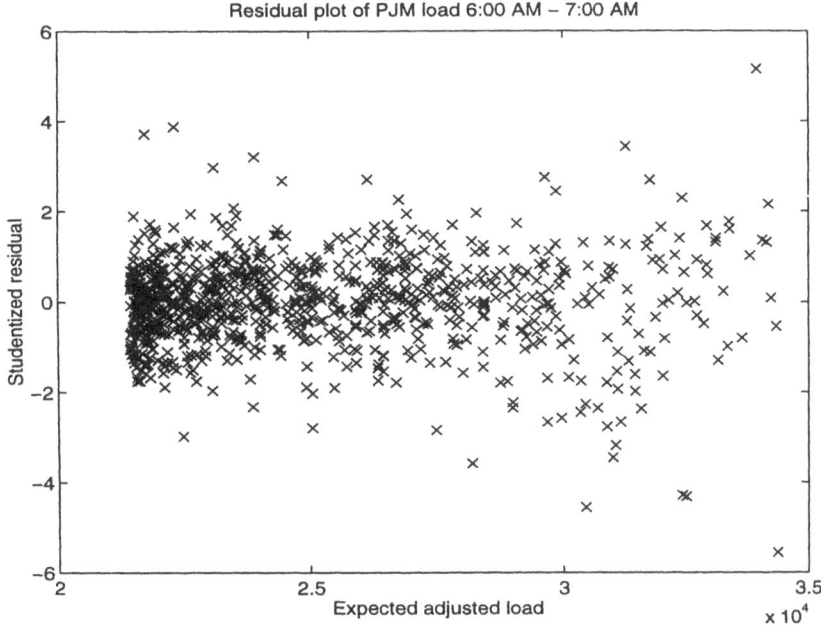

Figure B.22: Residual plot for 6:00 AM - 7:00 AM.

Studentized Residual	Day		Date	Adjusted Load Actual	Expected	High	Low
−3.449	8	Sat	1/8/94	27506	31006	33	18
−3.174	14	Fri	1/14/94	27383	31063	41	17
5.157	16	Sun	1/16/94	39604	33939	17	4
−5.556	18	Tue	1/18/94	28575	34368	34	2
3.885	217	Fri	8/5/94	27340	22304	91	60
3.722	241	Mon	8/29/94	26818	21732	89	60
−3.575	363	Thu	12/29/94	23161	28186	56	26
3.214	583	Sun	8/6/95	29412	23915	94	71
−4.558	725	Tue	12/26/95	25414	30461	32	22
−4.314	738	Mon	1/8/96	27786	32495	22	14
−4.284	739	Tue	1/9/96	27809	32418	26	14
3.450	750	Sat	1/20/96	31955	31270	62	19

Table B.53: List of cases with high residuals, 6:00 AM - 7:00 AM.

r_s	Day		Date	Prev. Day Low	Prev. Day High	Low	High	Next Day Low	Next Day High
−3.449	8	Sat	1/8/94	30	33	18	37	17	28
−3.174	14	Fri	1/14/94	36	41	17	39	6	17
5.157	16	Sun	1/16/94	6	17	4	15	14	34
−5.556	18	Tue	1/18/94	14	34	2	33	−5	6
3.885	217	Fri	8/5/94	76	91	60	86	55	77
3.722	241	Mon	8/29/94	70	89	60	79	59	79
−3.575	363	Thu	12/29/94	31	56	26	46	22	38
3.214	583	Sun	8/6/95	77	94	71	80	68	81
−4.558	725	Tue	12/26/95	27	32	22	31	21	31
−4.314	738	Mon	1/8/96	12	22	14	26	14	29
−4.284	739	Tue	1/9/96	14	26	14	29	20	32
3.450	750	Sat	1/20/96	22	62	19	29	22	32

Table B.54: List of cases with high residuals, 6:00 AM - 7:00 AM.

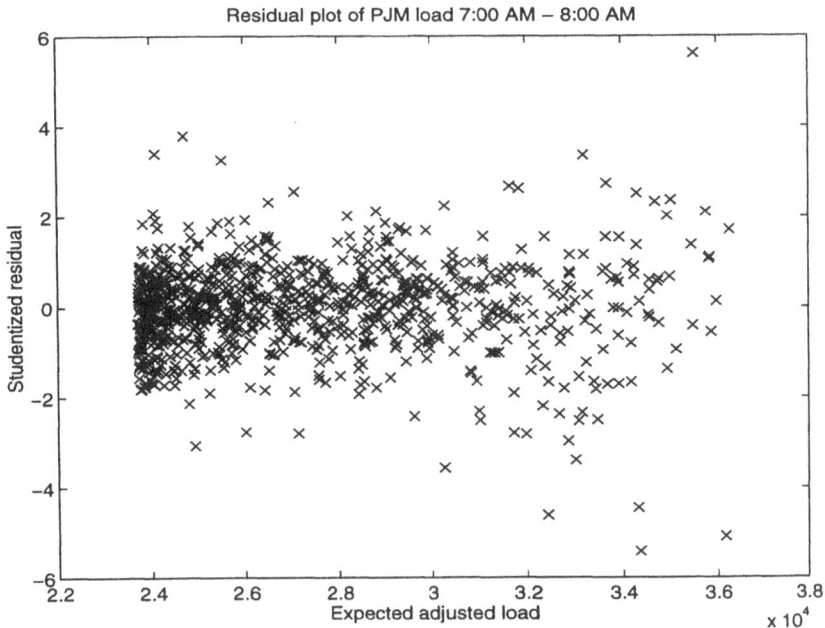

Figure B.23: Residual plot for 7:00 AM - 8:00 AM.

Studentized Residual	Day		Date	Adjusted Load Actual	Expected	High	Low
−3.387	8	Sat	1/8/94	29315	33002	33	18
5.627	16	Sun	1/16/94	42055	35518	17	4
−5.081	18	Tue	1/18/94	30528	36148	34	2
3.788	217	Fri	8/5/94	29914	24706	91	60
3.400	241	Mon	8/29/94	29111	24100	89	60
−3.558	363	Thu	12/29/94	25072	30267	56	26
−3.072	549	Mon	7/3/95	21441	24935	84	64
−4.615	725	Tue	12/26/95	27021	32420	32	22
−5.417	738	Mon	1/8/96	28081	34357	22	14
−4.451	739	Tue	1/9/96	29204	34315	26	14
3.369	750	Sat	1/20/96	34211	33196	62	19
3.255	872	Tue	5/21/96	29950	25526	94	60

Table B.55: List of cases with high residuals, 7:00 AM - 8:00 AM.

r_s	Day		Date	Prev. Day				Next Day	
				Low	High	Low	High	Low	High
−3.387	8	Sat	1/8/94	30	33	18	37	17	28
5.627	16	Sun	1/16/94	6	17	4	15	14	34
−5.081	18	Tue	1/18/94	14	34	2	33	−5	6
3.788	217	Fri	8/5/94	76	91	60	86	55	77
3.400	241	Mon	8/29/94	70	89	60	79	59	79
−3.558	363	Thu	12/29/94	31	56	26	46	22	38
−3.072	549	Mon	7/3/95	70	84	64	83	65	86
−4.615	725	Tue	12/26/95	27	32	22	31	21	31
−5.417	738	Mon	1/8/96	12	22	14	26	14	29
−4.451	739	Tue	1/9/96	14	26	14	29	20	32
3.369	750	Sat	1/20/96	22	62	19	29	22	32
3.255	872	Tue	5/21/96	68	94	60	91	55	80

Table B.56: List of cases with high residuals, 7:00 AM - 8:00 AM.

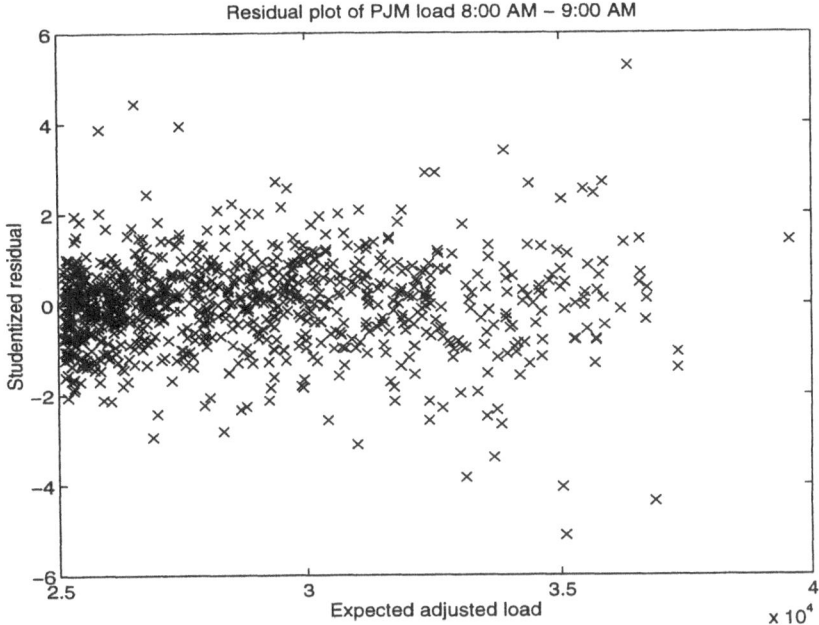

Figure B.24: Residual plot for 8:00 AM - 9:00 AM.

Studentized Residual	Day		Date	Adjusted Load Actual	Expected	High	Low
−3.384	8	Sat	1/8/94	30148	33691	33	18
5.268	16	Sun	1/16/94	42273	36347	17	4
−4.346	18	Tue	1/18/94	32190	36864	34	2
4.434	217	Fri	8/5/94	32386	26534	91	60
3.871	241	Mon	8/29/94	31375	25823	89	60
−3.090	363	Thu	12/29/94	26502	30994	56	26
−3.830	725	Tue	12/26/95	28779	33136	32	22
−5.108	738	Mon	1/8/96	29374	35081	22	14
−4.037	739	Tue	1/9/96	30556	35023	26	14
3.407	750	Sat	1/20/96	34902	33911	62	19
3.943	872	Tue	5/21/96	32544	27439	94	60

Table B.57: List of cases with high residuals, 8:00 AM - 9:00 AM.

r_s	Day		Date	Prev. Day				Next Day	
				Low	High	Low	High	Low	High
−3.384	8	Sat	1/8/94	30	33	18	37	17	28
5.268	16	Sun	1/16/94	6	17	4	15	14	34
−4.346	18	Tue	1/18/94	14	34	2	33	−5	6
4.434	217	Fri	8/5/94	76	91	60	86	55	77
3.871	241	Mon	8/29/94	70	89	60	79	59	79
−3.090	363	Thu	12/29/94	31	56	26	46	22	38
−3.830	725	Tue	12/26/95	27	32	22	31	21	31
−5.108	738	Mon	1/8/96	12	22	14	26	14	29
−4.037	739	Tue	1/9/96	14	26	14	29	20	32
3.407	750	Sat	1/20/96	22	62	19	29	22	32
3.943	872	Tue	5/21/96	68	94	60	91	55	80

Table B.58: List of cases with high residuals, 8:00 AM - 9:00 AM.

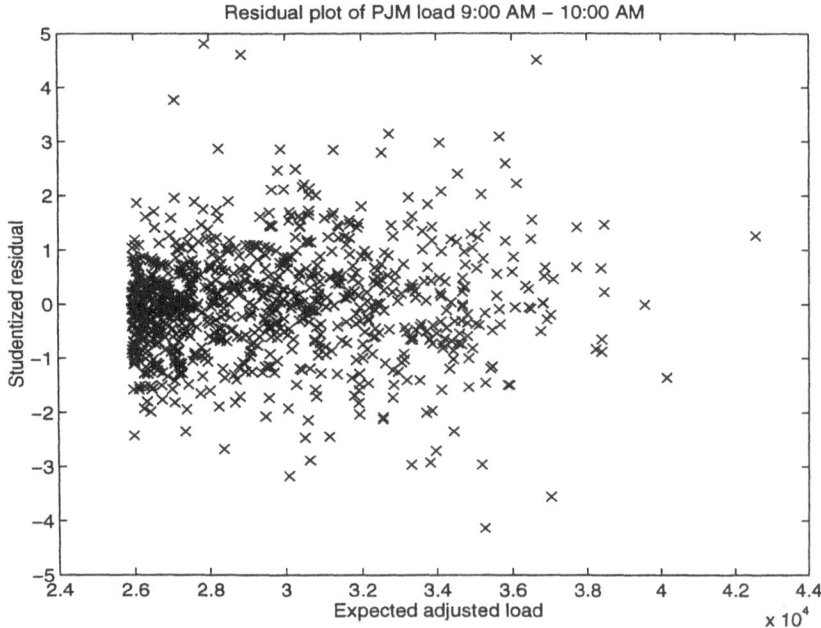

Figure B.25: Residual plot for 9:00 AM - 10:00 AM.

Studentized Residual	Day		Date	Adjusted Load Actual	Expected	High	Low
4.511	16	Sun	1/16/94	41805	36655	17	4
−3.535	18	Tue	1/18/94	33132	37007	34	2
3.149	23	Sun	1/23/94	36324	32733	33	24
4.797	217	Fri	8/5/94	34272	27882	91	60
3.762	241	Mon	8/29/94	32699	27070	89	60
−3.169	610	Sat	9/2/95	27660	30092	90	69
3.097	737	Sun	1/7/96	39098	35665	20	12
−4.121	738	Mon	1/8/96	30609	35257	22	14
4.595	872	Tue	5/21/96	34796	28870	94	60

Table B.59: List of cases with high residuals, 9:00 AM - 10:00 AM.

r_s	Day	Date		Prev. Day Low	Prev. Day High	Low	High	Next Day Low	Next Day High
4.511	16	Sun	1/16/94	6	17	4	15	14	34
−3.535	18	Tue	1/18/94	14	34	2	33	−5	6
3.149	23	Sun	1/23/94	15	33	24	33	31	51
4.797	217	Fri	8/5/94	76	91	60	86	55	77
3.762	241	Mon	8/29/94	70	89	60	79	59	79
−3.169	610	Sat	9/2/95	75	90	69	83	64	86
3.097	737	Sun	1/7/96	10	20	12	22	14	26
−4.121	738	Mon	1/8/96	12	22	14	26	14	29
4.595	872	Tue	5/21/96	68	94	60	91	55	80

Table B.60: List of cases with high residuals, 9:00 AM - 10:00 AM.

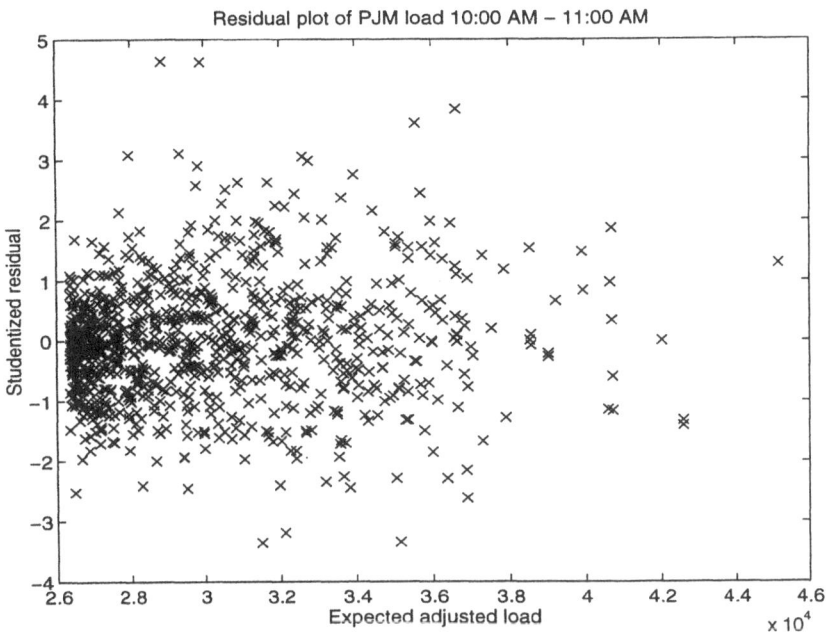

Figure B.26: Residual plot for 10:00 AM - 11:00 AM.

Studentized Residual	Day		Date	Actual	Expected	High	Low
				Adjusted Load			
3.842	16	Sun	1/16/94	41339	36610	17	4
3.057	23	Sun	1/23/94	36370	32618	33	24
4.626	217	Fri	8/5/94	35523	28855	91	60
3.079	241	Mon	8/29/94	33213	27958	89	60
-3.184	548	Sun	7/2/95	29127	32125	89	70
-3.350	610	Sat	9/2/95	28672	31503	90	69
3.610	737	Sun	1/7/96	39894	35568	20	12
-3.331	738	Mon	1/8/96	31107	35138	22	14
4.613	872	Tue	5/21/96	36295	29913	94	60
3.103	906	Mon	6/24/96	32767	29337	81	63

Table B.61: List of cases with high residuals, 10:00 AM - 11:00 AM.

r_s	Day		Date	Prev. Day Low	High	Low	High	Next Day Low	High
3.842	16	Sun	1/16/94	6	17	4	15	14	34
3.057	23	Sun	1/23/94	15	33	24	33	31	51
4.626	217	Fri	8/5/94	76	91	60	86	55	77
3.079	241	Mon	8/29/94	70	89	60	79	59	79
−3.184	548	Sun	7/2/95	71	89	70	84	64	83
−3.350	610	Sat	9/2/95	75	90	69	83	64	86
3.610	737	Sun	1/7/96	10	20	12	22	14	26
−3.331	738	Mon	1/8/96	12	22	14	26	14	29
4.613	872	Tue	5/21/96	68	94	60	91	55	80
3.103	906	Mon	6/24/96	68	81	63	84	68	87

Table B.62: List of cases with high residuals, 10:00 AM - 11:00 AM.

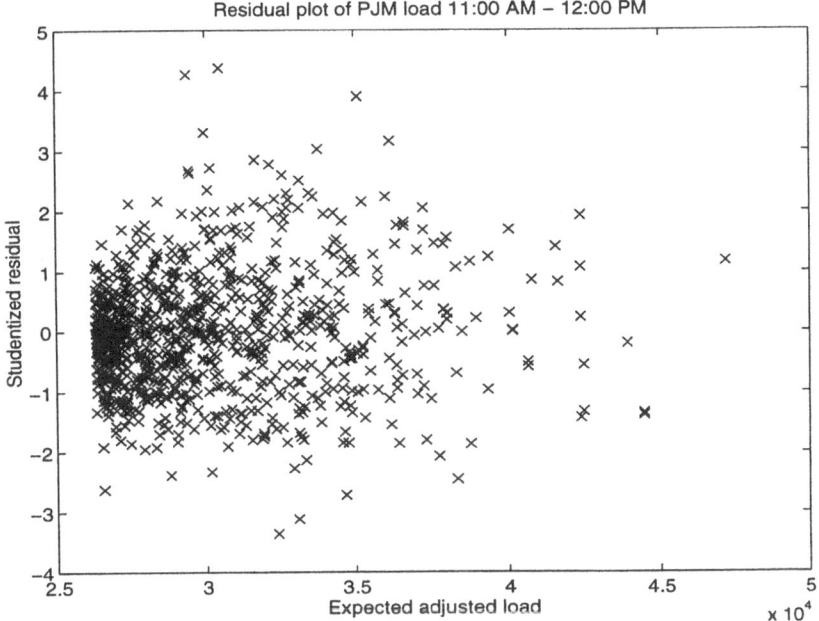

Figure B.27: Residual plot for 11:00 AM - 12:00 PM.

Studentized Residual	Day		Date	Adjusted Load Actual	Adjusted Load Expected	High	Low
3.168	16	Sun	1/16/94	40441	36162	17	4
4.261	217	Fri	8/5/94	36136	29368	91	60
3.026	226	Sun	8/14/94	38363	33788	92	69
−3.113	548	Sun	7/2/95	29918	33114	89	70
−3.359	610	Sat	9/2/95	29295	32423	90	69
3.901	737	Sun	1/7/96	40197	35098	20	12
4.371	872	Tue	5/21/96	37109	30491	94	60
3.303	906	Mon	6/24/96	33967	29960	81	63

Table B.63: List of cases with high residuals, 11:00 AM - 12:00 PM.

r_s	Day		Date	Prev. Day				Next Day	
				Low	High	Low	High	Low	High
3.168	16	Sun	1/16/94	6	17	4	15	14	34
4.261	217	Fri	8/5/94	76	91	60	86	55	77
3.026	226	Sun	8/14/94	75	92	69	90	62	75
−3.113	548	Sun	7/2/95	71	89	70	84	64	83
−3.359	610	Sat	9/2/95	75	90	69	83	64	86
3.901	737	Sun	1/7/96	10	20	12	22	14	26
4.371	872	Tue	5/21/96	68	94	60	91	55	80
3.303	906	Mon	6/24/96	68	81	63	84	68	87

Table B.64: List of cases with high residuals, 11:00 AM - 12:00 PM.

APPENDIX C
DERIVATION OF FORMULAS FOR TRUNCATED RANDOM VARIABLES

C.1   Truncated Normal Distributions

Throughout this discussion, we will assume that X is a normally distributed random variable with mean m and standard deviation σ. Z is a truncated normal variable with a minimum Z_{min} and a maximum Z_{max}; Z may be generated from a normal random variable X by:

$$Z = \begin{cases} Z_{min} & X \leq Z_{min} \\ X & Z_{min} < X < Z_{max} \\ Z_{max} & X \geq Z_{max} \end{cases} \tag{C.1}$$

Therefore, Z is continuously distributed between Z_{min} and Z_{max} but also has a finite probability of being equal to Z_{min} or Z_{max}. We are interested in finding the mean and variance of Z, as well as computing the expected value of XZ, as these are all quantities we will need when calculating the expected profit.

We begin by noting that the density of X is [42]:

$$d(x) = \frac{1}{\sigma\sqrt{2\pi}} e^{-\frac{(x-m)^2}{2\sigma^2}} \tag{C.2}$$

$d(x)$ also expresses the density of Z between Z_{min} and Z_{max}. In subsequent calculations, we will make use of the following integrals:

$$\int_{-\infty}^{a} e^{-\frac{x^2}{2\sigma^2}} dx = \sigma\sqrt{2\pi}\left(\frac{1}{2} + \frac{1}{2}\operatorname{erf}\left(\frac{a}{\sigma\sqrt{2}}\right)\right) \tag{C.3}$$

$$\int_a^\infty e^{-\frac{x^2}{2\sigma^2}} dx = \sigma\sqrt{2\pi}\left(\frac{1}{2} - \frac{1}{2}\operatorname{erf}\left(\frac{a}{\sigma\sqrt{2}}\right)\right) \tag{C.4}$$

$$\int_{-\infty}^a x e^{-\frac{x^2}{2\sigma^2}} dx = -\sigma^2 e^{-\frac{a^2}{2\sigma^2}} \tag{C.5}$$

$$\int_a^\infty x e^{-\frac{x^2}{2\sigma^2}} dx = \sigma^2 e^{-\frac{a^2}{2\sigma^2}} \tag{C.6}$$

$$\int_{-\infty}^a x^2 e^{-\frac{x^2}{2\sigma^2}} dx = -a\sigma^2 e^{-\frac{a^2}{2\sigma^2}} + \sigma^3\sqrt{2\pi}\left(\frac{1}{2} + \frac{1}{2}\operatorname{erf}\left(\frac{a}{\sigma\sqrt{2}}\right)\right) \tag{C.7}$$

$$\int_a^\infty x^2 e^{-\frac{x^2}{2\sigma^2}} dx = a\sigma^2 e^{-\frac{a^2}{2\sigma^2}} + \sigma^3\sqrt{2\pi}\left(\frac{1}{2} - \frac{1}{2}\operatorname{erf}\left(\frac{a}{\sigma\sqrt{2}}\right)\right) \tag{C.8}$$

The error function, $\operatorname{erf}(x)$, is defined as [42]:

$$\operatorname{erf}(x) = \frac{1}{\sqrt{\pi}}\int_{-x}^x e^{-t^2} dt \tag{C.9}$$

A normally distributed random variable with mean 0 and standard deviation $1/\sqrt{2}$ has probability $\operatorname{erf}(x)$ of lying between $-x$ and x, where $x \geq 0$. Note that $\operatorname{erf}(-x) = -\operatorname{erf}(x)$.

We also show that the density function integrates to 1 over the range of real numbers; i.e.:

$$\int_{-\infty}^\infty d(x)dx = 1 \tag{C.10}$$

To show this, we first define A as:

$$A = \int_{-\infty}^\infty e^{-\frac{(x-m)^2}{2\sigma^2}} dx \tag{C.11}$$

It is easy to see that this definition is the same as:

$$A = \int_{-\infty}^\infty e^{-\frac{x^2}{2\sigma^2}} dx \tag{C.12}$$

Then:

$$A^2 = \int_{-\infty}^\infty \int_{-\infty}^\infty e^{-\frac{x^2+y^2}{2\sigma^2}} dy\, dx \tag{C.13}$$

The integral can be transformed to polar coordinates, giving:

$$A^2 = \int_0^{2\pi} \int_0^\infty e^{-\frac{r^2}{2\sigma^2}} r\, dr\, d\theta \tag{C.14}$$

Performing the first integration, we have:

$$A^2 = \int_0^{2\pi} \sigma^2 d\theta \tag{C.15}$$

After performing the second integration and taking the square root of the answer, we have the value of A:

$$A = \sigma\sqrt{2\pi} \tag{C.16}$$

We have therefore demonstrated that the density $d(x)$ integrates to 1, since:

$$\int_{-\infty}^{\infty} d(x)dx = \frac{A}{\sigma\sqrt{2\pi}} = 1 \tag{C.17}$$

The preceding derivation is based on [42].

Next, we wish to show that the expected value of X is m. The expected value of a continuous random variable is found by [42]:

$$E(X) = \int_{-\infty}^{\infty} x\, d(x)\, dx \tag{C.18}$$

Substituting for $d(x)$:

$$E(X) = \frac{1}{\sigma\sqrt{2\pi}} \int_{-\infty}^{\infty} x e^{-\frac{(x-m)^2}{2\sigma^2}} dx \tag{C.19}$$

This integral is equivalent to:

$$E(X) = \frac{1}{\sigma\sqrt{2\pi}} \int_{-\infty}^{\infty} (x+m)e^{-\frac{x^2}{2\sigma^2}} dx \tag{C.20}$$

which may be broken into two separate integrals:

$$E(X) = \frac{1}{\sigma\sqrt{2\pi}} \int_{-\infty}^{\infty} x e^{-\frac{x^2}{2\sigma^2}} dx + \frac{m}{\sigma\sqrt{2\pi}} \int_{-\infty}^{\infty} e^{-\frac{x^2}{2\sigma^2}} dx \tag{C.21}$$

From equation (C.5), the first integral is zero, while the second integral is the same as A in equation (C.12), which we just showed is equal to $\sigma\sqrt{2\pi}$. Therefore, the expected value of X is indeed equal to m.

We will now determine the expected value of Z. The mean of Z may be calculated as:

$$\begin{aligned} E(Z) &= \frac{1}{\sigma\sqrt{2\pi}} \int_{Z_{min}}^{Z_{max}} x e^{-\frac{(x-m)^2}{2\sigma^2}} dx \\ &\quad + Z_{min}P(Z=Z_{min}) + Z_{max}P(Z=Z_{max}) \end{aligned} \tag{C.22}$$

where the notation $P(E)$ refers to the probability of event E occurring. These probabilities may be expressed as integrals of the density function, yielding:

$$
\begin{aligned}
E(Z) \;=\; & \frac{1}{\sigma\sqrt{2\pi}}\left[\int_{-\infty}^{\infty} x e^{-\frac{(x-m)^2}{2\sigma^2}}\,dx - \int_{-\infty}^{Z_{min}} x e^{-\frac{(x-m)^2}{2\sigma^2}}\,dx \right.\\
& -\int_{Z_{max}}^{\infty} x e^{-\frac{(x-m)^2}{2\sigma^2}}\,dx \\
& \left. + Z_{min}\int_{-\infty}^{Z_{min}} e^{-\frac{(x-m)^2}{2\sigma^2}}\,dx + Z_{max}\int_{Z_{max}}^{\infty} e^{-\frac{(x-m)^2}{2\sigma^2}}\,dx \right] \quad \text{(C.23)}
\end{aligned}
$$

where the combination of the first three integrals is equal to the single integral in equation (C.22). From equation (C.19), we see that the first integral in equation (C.23) is equal to $m\sigma\sqrt{2\pi}$. After making this substitution and also replacing x with $x+m$ in the remaining integrals, we obtain:

$$
\begin{aligned}
E(Z) \;=\; & m + \frac{1}{\sigma\sqrt{2\pi}}\left[-\int_{-\infty}^{Z_{min}-m} (x+m)e^{-\frac{x^2}{2\sigma^2}}\,dx \right.\\
& -\int_{Z_{max}-m}^{\infty} (x+m)e^{-\frac{x^2}{2\sigma^2}}\,dx \\
& \left. + Z_{min}\int_{-\infty}^{Z_{min}-m} e^{-\frac{x^2}{2\sigma^2}}\,dx + Z_{max}\int_{Z_{max}-m}^{\infty} e^{-\frac{x^2}{2\sigma^2}}\,dx \right] \quad \text{(C.24)}
\end{aligned}
$$

This can be rewritten as:

$$
\begin{aligned}
E(Z) \;=\; & m + \frac{1}{\sigma\sqrt{2\pi}}\left[-\int_{-\infty}^{Z_{min}-m} x e^{-\frac{x^2}{2\sigma^2}}\,dx - \int_{Z_{max}-m}^{\infty} x e^{-\frac{x^2}{2\sigma^2}}\,dx \right.\\
& \left. + (Z_{min}-m)\int_{-\infty}^{Z_{min}-m} e^{-\frac{x^2}{2\sigma^2}}\,dx + (Z_{max}-m)\int_{Z_{max}-m}^{\infty} e^{-\frac{x^2}{2\sigma^2}}\,dx \right]
\end{aligned}
$$

$$\text{(C.25)}$$

We can use equations (C.3) through (C.6) to evaluate the integrals. Upon doing so, we have:

$$
\begin{aligned}
E(Z) \;=\; & m + \frac{1}{\sigma\sqrt{2\pi}}\left[\sigma^2 e^{-\frac{(Z_{min}-m)^2}{2\sigma^2}} - \sigma^2 e^{-\frac{(Z_{max}-m)^2}{2\sigma^2}} \right.\\
& + (Z_{min}-m)(\sigma\sqrt{2\pi})\left(\frac{1}{2} + \frac{1}{2}\,\mathrm{erf}\left(\frac{Z_{min}-m}{\sigma\sqrt{2}}\right)\right) \\
& \left. + (Z_{max}-m)(\sigma\sqrt{2\pi})\left(\frac{1}{2} - \frac{1}{2}\,\mathrm{erf}\left(\frac{Z_{max}-m}{\sigma\sqrt{2}}\right)\right) \right]
\end{aligned}
$$

$$\text{(C.26)}$$

This equation can be simplified, giving our answer for the mean of Z:

$$
\begin{aligned}
E(Z) \;=\; & m + \frac{\sigma}{\sqrt{2\pi}} \left(e^{-\frac{(Z_{min}-m)^2}{2\sigma^2}} - e^{-\frac{(Z_{max}-m)^2}{2\sigma^2}} \right) + (Z_{min}-m)\left(\frac{1}{2}\right. \\
& + \frac{1}{2}\,\text{erf}\left(\frac{Z_{min}-m}{\sigma\sqrt{2}}\right) \Bigg) + (Z_{max}-m)\left(\frac{1}{2} - \frac{1}{2}\,\text{erf}\left(\frac{Z_{max}-m}{\sigma\sqrt{2}}\right)\right)
\end{aligned}
\tag{C.27}
$$

We will subsequently find it useful to define:

$$
\begin{aligned}
L_{CF} \;=\; & \frac{\sigma}{\sqrt{2\pi}} \left(e^{-\frac{(Z_{min}-m)^2}{2\sigma^2}} - e^{-\frac{(Z_{max}-m)^2}{2\sigma^2}} \right) + (Z_{min}-m)\left(\frac{1}{2}\right. \\
& + \frac{1}{2}\,\text{erf}\left(\frac{Z_{min}-m}{\sigma\sqrt{2}}\right) \Bigg) + (Z_{max}-m)\left(\frac{1}{2} - \frac{1}{2}\,\text{erf}\left(\frac{Z_{max}-m}{\sigma\sqrt{2}}\right)\right)
\end{aligned}
\tag{C.28}
$$

so that we can write:

$$
E(Z) = m + L_{CF}
\tag{C.29}
$$

L_{CF} is a "correction factor" to account for the change in mean of Z due to the presence of upper and lower limits.

Next, we wish to find the variance of Z. As before, we prepare for this calculation by first evaluating the variance of X. Recall that variance is defined by [42]:

$$
\text{var}(X) = E((X - E(X))^2)
\tag{C.30}
$$

For a normally distributed variable X, this expectation is computed by the integral:

$$
\text{var}(X) = \int_{-\infty}^{\infty} (x-m)^2 d(x)\,dx = \frac{1}{\sigma\sqrt{2\pi}} \int_{-\infty}^{\infty} (x-m)^2 e^{-\frac{(x-m)^2}{2\sigma^2}}\,dx
\tag{C.31}
$$

By applying the change of variable $y = x - m$ in the integrand, we have:

$$
\text{var}(X) = \frac{1}{\sigma\sqrt{2\pi}} \int_{-\infty}^{\infty} y^2 e^{-\frac{y^2}{2\sigma^2}}\,dy
\tag{C.32}
$$

We can now apply equation (C.7) to this integral. The first term in the result is zero, so the remaining term evaluates to:

$$
\text{var}(X) = \lim_{a \to \infty} \sigma^2 \left(\frac{1}{2} + \frac{1}{2}\,\text{erf}\left(\frac{a}{\sigma\sqrt{2}}\right) \right)
\tag{C.33}
$$

Since $\text{erf}(a) \to 1$ as $a \to \infty$ (as shown by equation (C.12) with $\sigma = 1/\sqrt{2}$), the variance of X is verified to be σ^2, as expected.

The procedure for finding the variance of Z is similar; however, we must remember that the expected value of Z is not in general m. The equation for the variance of Z is similar to equation (C.22), which we used earlier to find the mean:

$$
\begin{aligned}
\text{var}(Z) = {} & \frac{1}{\sigma\sqrt{2\pi}} \int_{Z_{min}}^{Z_{max}} (x - m - L_{CF})^2 e^{-\frac{(x-m)^2}{2\sigma^2}} \, dx \\
& + (Z_{min} - m - L_{CF})^2 P(Z = Z_{min}) \\
& + (Z_{max} - m - L_{CF})^2 P(Z = Z_{max})
\end{aligned}
\tag{C.34}
$$

The first integral may be rewritten as:

$$
\begin{aligned}
\text{var}(Z) = {} & \frac{1}{\sigma\sqrt{2\pi}} \left[\int_{-\infty}^{\infty} (x - m - L_{CF})^2 e^{-\frac{(x-m)^2}{2\sigma^2}} \, dx \right. \\
& - \int_{-\infty}^{Z_{min}} (x - m - L_{CF})^2 e^{-\frac{(x-m)^2}{2\sigma^2}} \, dx \\
& \left. - \int_{Z_{max}}^{\infty} (x - m - L_{CF})^2 e^{-\frac{(x-m)^2}{2\sigma^2}} \, dx \right] \\
& + (Z_{min} - m - L_{CF})^2 P(Z = Z_{min}) \\
& + (Z_{max} - m - L_{CF})^2 P(Z = Z_{max})
\end{aligned}
\tag{C.35}
$$

We now can expand the quadratic expression in the first integral, while replacing x with $x + m$ in the integrands of the other two integrals, arriving at:

$$
\begin{aligned}
\text{var}(Z) = {} & \frac{1}{\sigma\sqrt{2\pi}} \left[\int_{-\infty}^{\infty} \left((x - m)^2 - 2(x - m)L_{CF} + L_{CF}^2 \right) e^{-\frac{(x-m)^2}{2\sigma^2}} \, dx \right. \\
& \left. - \int_{-\infty}^{Z_{min}-m} (x - L_{CF})^2 e^{-\frac{x^2}{2\sigma^2}} \, dx - \int_{Z_{max}-m}^{\infty} (x - L_{CF})^2 e^{-\frac{x^2}{2\sigma^2}} \, dx \right] \\
& + (Z_{min} - m - L_{CF})^2 P(Z = Z_{min}) \\
& + (Z_{max} - m - L_{CF})^2 P(Z = Z_{max})
\end{aligned}
\tag{C.36}
$$

Next, we separate the first integral into three separate integrals, while expanding the squared term in the last two integrals:

$$
\begin{aligned}
\text{var}(Z) = {} & \frac{1}{\sigma\sqrt{2\pi}} \left[\int_{-\infty}^{\infty} (x - m)^2 e^{-\frac{(x-m)^2}{2\sigma^2}} \, dx - 2L_{CF} \int_{-\infty}^{\infty} x e^{-\frac{(x-m)^2}{2\sigma^2}} \, dx \right. \\
& + (2mL_{CF} + L_{CF}^2) \int_{-\infty}^{\infty} e^{-\frac{(x-m)^2}{2\sigma^2}} \, dx
\end{aligned}
$$

$$- \int_{-\infty}^{Z_{min}-m} (x^2 - 2L_{CF}x + L_{CF}^2)e^{-\frac{x^2}{2\sigma^2}}\,dx$$

$$- \int_{Z_{max}-m}^{\infty} (x^2 - 2L_{CF}x + L_{CF}^2)e^{-\frac{x^2}{2\sigma^2}}\,dx \Bigg]$$

$$+ (Z_{min} - m - L_{CF})^2 P(Z = Z_{min})$$

$$+ (Z_{max} - m - L_{CF})^2 P(Z = Z_{max}) \tag{C.37}$$

If we include the constant factor $1/\sigma\sqrt{2\pi}$ in the integrand of the first three integrals, the first integral in equation (C.37) is the variance of X, which is σ^2. The second integral is the expectation of X, which we know is equal to m. Finally, the third integral is an integration of the density function over all real numbers, which we know must equal 1. Making these substitutions into equation (C.37) and expressing the two probabilities as integrals gives:

$$\mathrm{var}(Z) = \sigma^2 + L_{CF}^2 + \frac{1}{\sigma\sqrt{2\pi}} \Bigg[- \int_{-\infty}^{Z_{min}-m} x^2 e^{-\frac{x^2}{2\sigma^2}}\,dx$$

$$+ 2L_{CF} \int_{-\infty}^{Z_{min}-m} x e^{-\frac{x^2}{2\sigma^2}}\,dx - L_{CF}^2 \int_{-\infty}^{Z_{min}-m} e^{-\frac{x^2}{2\sigma^2}}\,dx$$

$$- \int_{Z_{max}-m}^{\infty} x^2 e^{-\frac{x^2}{2\sigma^2}}\,dx + 2L_{CF} \int_{Z_{max}-m}^{\infty} x e^{-\frac{x^2}{2\sigma^2}}\,dx$$

$$- L_{CF}^2 \int_{Z_{max}-m}^{\infty} e^{-\frac{x^2}{2\sigma^2}}\,dx$$

$$+ (Z_{min} - m - L_{CF})^2 \int_{-\infty}^{Z_{min}} e^{-\frac{(x-m)^2}{2\sigma^2}}\,dx$$

$$+ (Z_{max} - m - L_{CF})^2 \int_{Z_{max}}^{-\infty} e^{-\frac{(x-m)^2}{2\sigma^2}}\,dx \Bigg] \tag{C.38}$$

Next, we replace x with $x + m$ in the last two integrands to get:

$$\mathrm{var}(Z) = \sigma^2 + L_{CF}^2 + \frac{1}{\sigma\sqrt{2\pi}} \Bigg[- \int_{-\infty}^{Z_{min}-m} x^2 e^{-\frac{x^2}{2\sigma^2}}\,dx$$

$$+ 2L_{CF} \int_{-\infty}^{Z_{min}-m} x e^{-\frac{x^2}{2\sigma^2}}\,dx$$

$$+ ((Z_{min} - m)^2 - 2L_{CF}(Z_{min} - m)) \int_{-\infty}^{Z_{min}-m} e^{-\frac{x^2}{2\sigma^2}}\,dx$$

$$- \int_{Z_{max}-m}^{\infty} x^2 e^{-\frac{x^2}{2\sigma^2}}\,dx + 2L_{CF} \int_{Z_{max}-m}^{\infty} x e^{-\frac{x^2}{2\sigma^2}}\,dx$$

$$+ ((Z_{max} - m)^2 - 2L_{CF}(Z_{max} - m)) \int_{Z_{max}-m}^{\infty} e^{-\frac{x^2}{2\sigma^2}}\,dx \Bigg] \tag{C.39}$$

We can now apply equations (C.3) through (C.8) to perform the integrations:

$$
\begin{aligned}
\text{var}(Z) \;=\; & \sigma^2 + L_{CF}^2 + \frac{1}{\sigma\sqrt{2\pi}}\left[(Z_{min}-m)\sigma^2 e^{-\frac{(Z_{min}-m)^2}{2\sigma^2}}\right.\\
& -\sigma^3\sqrt{2\pi}\left(\frac{1}{2}+\frac{1}{2}\operatorname{erf}\left(\frac{Z_{min}-m}{\sigma\sqrt{2}}\right)\right) - 2L_{CF}\sigma^2 e^{-\frac{(Z_{min}-m)^2}{2\sigma^2}}\\
& +\sigma\sqrt{2\pi}((Z_{min}-m)^2 - 2L_{CF}(Z_{min}-m))\left(\frac{1}{2}\right.\\
& +\frac{1}{2}\operatorname{erf}\left(\frac{Z_{min}-m}{\sigma\sqrt{2}}\right)\bigg) - (Z_{max}-m)\sigma^2 e^{-\frac{(Z_{max}-m)^2}{2\sigma^2}}\\
& -\sigma^3\sqrt{2\pi}\left(\frac{1}{2}-\frac{1}{2}\operatorname{erf}\left(\frac{Z_{max}-m}{\sigma\sqrt{2}}\right)\right) + 2L_{CF}\sigma^2 e^{-\frac{(Z_{max}-m)^2}{2\sigma^2}}\\
& +\sigma\sqrt{2\pi}((Z_{max}-m)^2\\
& -2L_{CF}(Z_{max}-m))\left(\frac{1}{2}-\frac{1}{2}\operatorname{erf}\left(\frac{Z_{max}-m}{\sigma\sqrt{2}}\right)\right)\bigg]
\end{aligned}
\tag{C.40}
$$

This equation may be reduced to:

$$
\begin{aligned}
\text{var}(Z) \;=\; & \sigma^2 + L_{CF}^2 + \frac{\sigma}{\sqrt{2\pi}}\left((Z_{min}-m-2L_{CF})e^{-\frac{(Z_{min}-m)^2}{2\sigma^2}}\right.\\
& -(Z_{max}-m-2L_{CF})e^{-\frac{(Z_{max}-m)^2}{2\sigma^2}}\bigg)\\
& +\left((Z_{min}-m)^2 - 2L_{CF}(Z_{min}-m)-\sigma^2\right)\left(\frac{1}{2}\right.\\
& +\frac{1}{2}\operatorname{erf}\left(\frac{Z_{min}-m}{\sigma\sqrt{2}}\right)\bigg) + ((Z_{max}-m)^2 - 2L_{CF}(Z_{max}-m)\\
& -\sigma^2)\left(\frac{1}{2}-\frac{1}{2}\operatorname{erf}\left(\frac{Z_{max}-m}{\sigma\sqrt{2}}\right)\right)
\end{aligned}
\tag{C.41}
$$

Finally, we can add L_{CF}^2 to complete the square in the first factor of the last two terms, giving the result for the variance of Z:

$$
\begin{aligned}
\text{var}(Z) \;=\; & (\sigma^2 + L_{CF}^2)\left(\frac{1}{2}\operatorname{erf}\left(\frac{Z_{max}-m}{\sigma\sqrt{2}}\right)-\frac{1}{2}\operatorname{erf}\left(\frac{Z_{min}-m}{\sigma\sqrt{2}}\right)\right)\\
& +(Z_{min}-m-L_{CF})^2\left(\frac{1}{2}+\frac{1}{2}\operatorname{erf}\left(\frac{Z_{min}-m}{\sigma\sqrt{2}}\right)\right)\\
& +(Z_{max}-m-L_{CF})^2\left(\frac{1}{2}-\frac{1}{2}\operatorname{erf}\left(\frac{Z_{max}-m}{\sigma\sqrt{2}}\right)\right)\\
& +\frac{\sigma}{\sqrt{2\pi}}\left((Z_{min}-m-2L_{CF})e^{-\frac{(Z_{min}-m)^2}{2\sigma^2}}\right.\\
& -(Z_{max}-m-2L_{CF})e^{-\frac{(Z_{max}-m)^2}{2\sigma^2}}\bigg)
\end{aligned}
\tag{C.42}
$$

Lastly, we wish to compute the expected value of the product of a normally distributed random variable and a truncated version of that same variable. Mathematically, we wish to determine the expected value of XZ where X is normally distributed and Z is related to X by equation (C.1). This expected value is used to compute expected revenue for a generator since, as will be seen, the price is treated as a normal random variable while the power sold is a function of price, but with upper and lower limits.

The most straightforward way to find $E(XZ)$ is to integrate the product of XZ and its density; this is the same method that was used earlier in equations (C.22) and (C.34) to find the mean and variance of Z:

$$E(XZ) = \frac{1}{\sigma\sqrt{2\pi}} \left[\int_{Z_{min}}^{Z_{max}} x^2 e^{-\frac{(x-m)^2}{2\sigma^2}} dx + \int_{-\infty}^{Z_{min}} Z_{min} x e^{-\frac{(x-m)^2}{2\sigma^2}} dx \right.$$
$$\left. + \int_{Z_{max}}^{\infty} Z_{max} x e^{-\frac{(x-m)^2}{2\sigma^2}} dx \right] \tag{C.43}$$

As before, we begin the evaluation of this expression by breaking the first integral into a sum of three integrals:

$$E(XZ) = \frac{1}{\sigma\sqrt{2\pi}} \left[\int_{-\infty}^{\infty} x^2 e^{-\frac{(x-m)^2}{2\sigma^2}} dx - \int_{-\infty}^{Z_{min}} x^2 e^{-\frac{(x-m)^2}{2\sigma^2}} dx \right.$$
$$- \int_{Z_{max}}^{\infty} x^2 e^{-\frac{(x-m)^2}{2\sigma^2}} dx + \int_{-\infty}^{Z_{min}} Z_{min} x e^{-\frac{(x-m)^2}{2\sigma^2}} dx$$
$$\left. + \int_{Z_{max}}^{\infty} Z_{max} x e^{-\frac{(x-m)^2}{2\sigma^2}} dx \right] \tag{C.44}$$

The first integral is $\sigma\sqrt{2\pi}$ times the expected value of X^2, which was shown earlier to be $m^2 + \sigma^2$. After substituting x with $x + m$ in the remaining integrals, we have:

$$E(XZ) = m^2 + \sigma^2 + \frac{1}{\sigma\sqrt{2\pi}} \left[-\int_{-\infty}^{Z_{min}-m} (x+m)^2 e^{-\frac{x^2}{2\sigma^2}} dx \right.$$
$$- \int_{Z_{max}-m}^{\infty} (x+m)^2 e^{-\frac{x^2}{2\sigma^2}} dx + \int_{-\infty}^{Z_{min}-m} Z_{min}(x+m) e^{-\frac{x^2}{2\sigma^2}} dx$$
$$\left. + \int_{Z_{max}-m}^{\infty} Z_{max}(x+m) e^{-\frac{x^2}{2\sigma^2}} dx \right] \tag{C.45}$$

Expanding the squares and collecting terms gives:

$$E(XZ) = m^2 + \sigma^2 + \frac{1}{\sigma\sqrt{2\pi}} \left[-\int_{-\infty}^{Z_{min}-m} x^2 e^{-\frac{x^2}{2\sigma^2}} dx - \int_{Z_{max}-m}^{\infty} x^2 e^{-\frac{x^2}{2\sigma^2}} dx \right.$$

$$+ (Z_{min} - 2m) \int_{-\infty}^{Z_{min}-m} x e^{-\frac{x^2}{2\sigma^2}} dx$$

$$+ (Z_{max} - 2m) \int_{Z_{max}-m}^{\infty} x e^{-\frac{x^2}{2\sigma^2}} dx$$

$$+ (mZ_{min} - m^2) \int_{-\infty}^{Z_{min}-m} e^{-\frac{x^2}{2\sigma^2}} dx$$

$$\left. + (mZ_{max} - m^2) \int_{Z_{max}-m}^{\infty} e^{-\frac{x^2}{2\sigma^2}} dx \right] \tag{C.46}$$

Next, applying equations (C.3) to (C.8) to carry out the integrations:

$$E(XZ) = m^2 + \sigma^2 + \frac{1}{\sigma\sqrt{2\pi}} \left[(Z_{min} - m)\sigma^2 e^{-\frac{(Z_{min}-m)^2}{2\sigma^2}} - \sigma^3\sqrt{2\pi} \left(\frac{1}{2} \right. \right.$$

$$\left. + \frac{1}{2}\operatorname{erf}\left(\frac{Z_{min} - m}{\sigma\sqrt{2}} \right) \right) - (Z_{max} - m)\sigma^2 e^{-\frac{(Z_{max}-m)^2}{2\sigma^2}}$$

$$- \sigma^3\sqrt{2\pi} \left(\frac{1}{2} - \frac{1}{2}\operatorname{erf}\left(\frac{Z_{max} - m}{\sigma\sqrt{2}} \right) \right)$$

$$- (Z_{min} - 2m)\sigma^2 e^{-\frac{(Z_{min}-m)^2}{2\sigma^2}} + (Z_{max} - 2m)\sigma^2 e^{-\frac{(Z_{max}-m)^2}{2\sigma^2}}$$

$$+ (mZ_{min} - m^2)\sigma\sqrt{2\pi} \left(\frac{1}{2} + \frac{1}{2}\operatorname{erf}\left(\frac{Z_{min} - m}{\sigma\sqrt{2}} \right) \right)$$

$$\left. + (mZ_{max} - m^2)\sigma\sqrt{2\pi} \left(\frac{1}{2} - \frac{1}{2}\operatorname{erf}\left(\frac{Z_{max} - m}{\sigma\sqrt{2}} \right) \right) \right] \tag{C.47}$$

This equation simplifies to:

$$E(XZ) = m^2 + \sigma^2 + \frac{m\sigma}{\sqrt{2\pi}} \left(e^{-\frac{(Z_{min}-m)^2}{2\sigma^2}} - e^{-\frac{(Z_{max}-m)^2}{2\sigma^2}} \right)$$

$$+ (mZ_{min} - m^2 - \sigma^2) \left(\frac{1}{2} + \frac{1}{2}\operatorname{erf}\left(\frac{Z_{min} - m}{\sigma\sqrt{2}} \right) \right)$$

$$+ (mZ_{max} - m^2 - \sigma^2) \left(\frac{1}{2} - \frac{1}{2}\operatorname{erf}\left(\frac{Z_{max} - m}{\sigma\sqrt{2}} \right) \right) \tag{C.48}$$

C.2 Truncated Lognormal Distributions

We will now consider the effects of limits on the mean and variance of a random variable whose logarithm is normally distributed. Given a truncated normal random variable Z with mean m, standard deviation σ, and limits Z_{min} and Z_{max}, we wish to find the mean and variance of e^Z. We also want to determine the expected value of $e^X e^Z$, where Z is a truncated version of the normally distributed variable X.

We begin by determining the expected value of e^Z. As before, the mean of e^Z may be calculated as:

$$E(e^Z) = \frac{1}{\sigma\sqrt{2\pi}} \left[\int_{Z_{min}}^{Z_{max}} e^x e^{-\frac{(x-m)^2}{2\sigma^2}} dx \right.$$

$$\left. + e^{Z_{min}} P(Z = Z_{min}) + e^{Z_{max}} P(Z = Z_{max}) \right] \tag{C.49}$$

We can use equation (A.4) to substitute for the first integral in equation (C.2); by making this substitution and also substituting integrals for the probability expressions, we obtain:

$$E(e^Z) = \frac{1}{\sigma\sqrt{2\pi}} \left[e^{m+\frac{1}{2}\sigma^2} \int_{Z_{min}}^{Z_{max}} e^{-\frac{(x-(m+\sigma^2))^2}{2\sigma^2}} dx \right.$$

$$\left. + e^{Z_{min}} \int_{-\infty}^{Z_{min}} e^{-\frac{(x-m)^2}{2\sigma^2}} dx + e^{Z_{max}} \int_{Z_{max}}^{\infty} e^{-\frac{(x-m)^2}{2\sigma^2}} dx \right] \tag{C.50}$$

By substituting x with $x + m + \sigma^2$ in the first integral and x with $x + m$ in the remaining integrals, this equation can be rewritten as:

$$E(e^Z) = \frac{1}{\sigma\sqrt{2\pi}} \left[e^{m+\frac{1}{2}\sigma^2} \int_{Z_{min}-m-\sigma^2}^{Z_{max}-m-\sigma^2} e^{-\frac{x^2}{2\sigma^2}} dx \right.$$

$$\left. + e^{Z_{min}} \int_{-\infty}^{Z_{min}-m} e^{-\frac{x^2}{2\sigma^2}} dx + e^{Z_{max}} \int_{Z_{max}-m}^{\infty} e^{-\frac{x^2}{2\sigma^2}} dx \right] \tag{C.51}$$

By applying the integral identity:

$$\int_a^b f(x)\, dx = \int_{-\infty}^b f(x)\, dx - \int_{-\infty}^a f(x)\, dx \tag{C.52}$$

we can use equation (C.3) to evaluate the integrals, as before. Upon doing so, we have:

$$E(e^Z) = \frac{1}{\sigma\sqrt{2\pi}} \left[\frac{1}{2}(\sigma\sqrt{2\pi}) e^{m+\frac{1}{2}\sigma^2} \left(\mathrm{erf}\left(\frac{Z_{max} - m - \sigma^2}{\sigma\sqrt{2}} \right) \right. \right.$$

$$\left. - \mathrm{erf}\left(\frac{Z_{min} - m - \sigma^2}{\sigma\sqrt{2}} \right) \right)$$

$$+ e^{Z_{min}} (\sigma\sqrt{2\pi}) \left(\frac{1}{2} + \frac{1}{2}\mathrm{erf}\left(\frac{Z_{min} - m}{\sigma\sqrt{2}} \right) \right)$$

$$\left. + e^{Z_{max}} (\sigma\sqrt{2\pi}) \left(\frac{1}{2} - \frac{1}{2}\mathrm{erf}\left(\frac{Z_{max} - m}{\sigma\sqrt{2}} \right) \right) \right]$$

$$\tag{C.53}$$

The mean of e^Z is therefore:

$$
\begin{aligned}
E(e^Z) \;=\;& \frac{1}{2}e^{m+\frac{1}{2}\sigma^2}\left(\operatorname{erf}\left(\frac{Z_{max}-m-\sigma^2}{\sigma\sqrt{2}}\right)-\operatorname{erf}\left(\frac{Z_{min}-m-\sigma^2}{\sigma\sqrt{2}}\right)\right)\\
&+e^{Z_{min}}\left(\frac{1}{2}+\frac{1}{2}\operatorname{erf}\left(\frac{Z_{min}-m}{\sigma\sqrt{2}}\right)\right)\\
&+e^{Z_{max}}\left(\frac{1}{2}-\frac{1}{2}\operatorname{erf}\left(\frac{Z_{max}-m}{\sigma\sqrt{2}}\right)\right)
\end{aligned}
\tag{C.54}
$$

We will subsequently use the notation:

$$
E(e^Z)=C_{CF}m+L_{CF}
\tag{C.55}
$$

with the quantities C_{CF} and L_{CF} defined by:

$$
C_{CF}=\frac{1}{2}e^{\frac{1}{2}\sigma^2}\left(\operatorname{erf}\left(\frac{Z_{max}-m-\sigma^2}{\sigma\sqrt{2}}\right)-\operatorname{erf}\left(\frac{Z_{min}-m-\sigma^2}{\sigma\sqrt{2}}\right)\right)
\tag{C.56}
$$

$$
\begin{aligned}
L_{CF} \;=\;& e^{Z_{min}}\left(\frac{1}{2}+\frac{1}{2}\operatorname{erf}\left(\frac{Z_{min}-m}{\sigma\sqrt{2}}\right)\right)\\
&+e^{Z_{max}}\left(\frac{1}{2}-\frac{1}{2}\operatorname{erf}\left(\frac{Z_{max}-m}{\sigma\sqrt{2}}\right)\right)
\end{aligned}
\tag{C.57}
$$

As with the normal distribution, C_{CF} and L_{CF} are quantities that described how the mean of e^Z is affected by the limits on Z.

Next, we wish to find the variance of e^Z. Following the same procedure as earlier:

$$
\begin{aligned}
\operatorname{var}(e^Z) \;=\;& \frac{1}{\sigma\sqrt{2\pi}}\int_{Z_{min}}^{Z_{max}}(e^x-C_{CF}m-L_{CF})^2 e^{-\frac{(x-m)^2}{2\sigma^2}}\,dx\\
&+(e^{Z_{min}}-C_{CF}m-L_{CF})^2 P(Z=Z_{min})\\
&+(e^{Z_{max}}-C_{CF}m-L_{CF})^2 P(Z=Z_{max})
\end{aligned}
\tag{C.58}
$$

Expanding the quadratic expression in the first integral:

$$
\begin{aligned}
\operatorname{var}(e^Z) \;=\;& \frac{1}{\sigma\sqrt{2\pi}}\int_{Z_{min}}^{Z_{max}} e^{2x} e^{-\frac{(x-m)^2}{2\sigma^2}}\,dx\\
&+e^{2Z_{min}}P(Z=Z_{min})+e^{2Z_{max}}P(Z=Z_{max})\\
&-2(C_{CF}m+L_{CF})\left[\frac{1}{\sigma\sqrt{2\pi}}\int_{Z_{min}}^{Z_{max}} e^x e^{-\frac{(x-m)^2}{2\sigma^2}}\,dx\right.\\
&\left.+e^{Z_{min}}P(Z=Z_{min})+e^{Z_{max}}P(Z=Z_{max})\right]
\end{aligned}
$$

$$+ (C_{CF}m + L_{CF})^2 \left[\frac{1}{\sigma\sqrt{2\pi}} \int_{Z_{min}}^{Z_{max}} e^{-\frac{(x-m)^2}{2\sigma^2}} dx \right.$$

$$\left. + P(Z = Z_{min}) + P(Z = Z_{max}) \right] \qquad \text{(C.59)}$$

The first bracketed expression in equation (C.59) is the mean of e^Z, while the second bracketed expression is the integral of the density function of Z, which is equal to 1. The first integral may be substituted according to equation (A.10). Equation (C.59) then simplifies to:

$$\text{var}(e^Z) = \frac{1}{\sigma\sqrt{2\pi}} \left[e^{2m+2\sigma^2} \int_{Z_{min}}^{Z_{max}} e^{-\frac{(x-(m+2\sigma^2))^2}{2\sigma^2}} dx \right.$$

$$+ e^{2Z_{min}} \int_{-\infty}^{Z_{min}} e^{-\frac{(x-m)^2}{2\sigma^2}} dx + e^{2Z_{max}} \int_{Z_{max}}^{\infty} e^{-\frac{(x-m)^2}{2\sigma^2}} dx \bigg]$$

$$- (C_{CF}m + L_{CF})^2 \qquad \text{(C.60)}$$

After changing the limits in the integrals:

$$\text{var}(e^Z) = \frac{1}{\sigma\sqrt{2\pi}} \left[e^{2m+2\sigma^2} \int_{Z_{min}-m-2\sigma^2}^{Z_{max}-m-2\sigma^2} e^{-\frac{x^2}{2\sigma^2}} dx \right.$$

$$+ e^{2Z_{min}} \int_{-\infty}^{Z_{min}-m} e^{-\frac{x^2}{2\sigma^2}} dx + e^{2Z_{max}} \int_{Z_{max}-m}^{\infty} e^{-\frac{x^2}{2\sigma^2}} dx \bigg]$$

$$- (C_{CF}m + L_{CF})^2 \qquad \text{(C.61)}$$

Finally, applying equation (C.3) to perform the integrations:

$$\text{var}(e^Z) = \frac{1}{2} e^{2m+2\sigma^2} \left(\text{erf}\left(\frac{Z_{max} - m - 2\sigma^2}{\sigma\sqrt{2}} \right) - \text{erf}\left(\frac{Z_{min} - m - 2\sigma^2}{\sigma\sqrt{2}} \right) \right)$$

$$+ e^{2Z_{min}} \left(\frac{1}{2} + \frac{1}{2}\,\text{erf}\left(\frac{Z_{min} - m}{\sigma\sqrt{2}} \right) \right)$$

$$+ e^{2Z_{max}} \left(\frac{1}{2} - \frac{1}{2}\,\text{erf}\left(\frac{Z_{max} - m}{\sigma\sqrt{2}} \right) \right) - (C_{CF}m + L_{CF})^2 \qquad \text{(C.62)}$$

Finally, we now compute the expected value of $e^X e^Z$, which is the product of a lognormally distributed random variable and a truncated version of that same variable. This quantity may be determined by evaluating the following expression:

$$E(e^{X+Z}) = \frac{1}{\sigma\sqrt{2\pi}} \left[\int_{Z_{min}}^{Z_{max}} e^{2x} e^{-\frac{(x-m)^2}{2\sigma^2}} dx + \int_{-\infty}^{Z_{min}} e^{Z_{min}} e^x e^{-\frac{(x-m)^2}{2\sigma^2}} dx \right.$$

$$+ \int_{Z_{max}}^{\infty} e^{Z_{max}} e^x e^{-\frac{(x-m)^2}{2\sigma^2}} dx \bigg] \qquad \text{(C.63)}$$

The integrals can be simplified by making use of equations (A.4) and (A.10):

$$E(e^{X+Z}) = \frac{1}{\sigma\sqrt{2\pi}} \left[e^{2m+2\sigma^2} \int_{Z_{min}}^{Z_{max}} e^{-\frac{(x-(m+2\sigma^2))^2}{2\sigma^2}} dx \right.$$

$$+ e^{Z_{min}} e^{m+\frac{1}{2}\sigma^2} \int_{-\infty}^{Z_{min}} e^{-\frac{(x-(m+\sigma^2))^2}{2\sigma^2}} dx$$

$$\left. + e^{Z_{max}} e^{m+\frac{1}{2}\sigma^2} \int_{Z_{max}}^{\infty} e^{-\frac{(x-(m+\sigma^2))^2}{2\sigma^2}} dx \right] \qquad \text{(C.64)}$$

This equation may be rewritten as:

$$E(e^{X+Z}) = \frac{1}{\sigma\sqrt{2\pi}} \left[e^{2m+2\sigma^2} \int_{Z_{min}-m-2\sigma^2}^{Z_{max}-m-2\sigma^2} e^{-\frac{x^2}{2\sigma^2}} dx \right.$$

$$+ e^{Z_{min}+m+\frac{1}{2}\sigma^2} \int_{-\infty}^{Z_{min}-m-\sigma^2} e^{-\frac{x^2}{2\sigma^2}} dx$$

$$\left. + e^{Z_{max}+m+\frac{1}{2}\sigma^2} \int_{Z_{max}-m-\sigma^2}^{\infty} e^{-\frac{x^2}{2\sigma^2}} dx \right] \qquad \text{(C.65)}$$

Next, using equations (C.3) and (C.4) to carry out the integrations:

$$E(e^{X+Z}) = \frac{1}{2} e^{2m+2\sigma^2} \left(\text{erf}\left(\frac{Z_{max}-m-2\sigma^2}{\sigma\sqrt{2}} \right) - \text{erf}\left(\frac{Z_{min}-m-2\sigma^2}{\sigma\sqrt{2}} \right) \right)$$

$$+ \frac{1}{2} e^{Z_{min}+m+\frac{1}{2}\sigma^2} \left(1 + \text{erf}\left(\frac{Z_{min}-m-\sigma^2}{\sigma\sqrt{2}} \right) \right)$$

$$+ \frac{1}{2} e^{Z_{max}+m+\frac{1}{2}\sigma^2} \left(1 - \text{erf}\left(\frac{Z_{max}-m-\sigma^2}{\sigma\sqrt{2}} \right) \right) \qquad \text{(C.66)}$$

We will also need higher powers of e^X and e^Z to find higher order moments of the distribution for the futures optimization. To derive these more general formulas, we begin by evaluating the following integral:

$$\int_a^b e^{kx} e^{-\frac{(x-m)^2}{2\sigma^2}} dx = \int_a^b e^{\frac{2k\sigma^2 x - (x^2 - 2mx + m^2)}{2\sigma^2}} dx \qquad \text{(C.67)}$$

Rearranging terms in the exponent:

$$\int_a^b e^{kx} e^{-\frac{(x-m)^2}{2\sigma^2}} dx = \int_a^b e^{\frac{-x^2+(2k\sigma^2+2m)x-m^2-2mk\sigma^2-k^2\sigma^4+2mk\sigma^2+k^2\sigma^4}{2\sigma^2}} dx \qquad \text{(C.68)}$$

The extra terms in the exponent are used to complete the square of $(x-(m+k\sigma^2))$:

$$\int_a^b e^{kx} e^{-\frac{(x-m)^2}{2\sigma^2}} dx = \int_a^b e^{\frac{-(x-(m+k\sigma^2))^2+2mk\sigma^2+k^2\sigma^4}{2\sigma^2}} dx \qquad \text{(C.69)}$$

This equation simplifies to:

$$\int_a^b e^{kx} e^{-\frac{(x-m)^2}{2\sigma^2}} \, dx = e^{mk+\frac{k^2}{2}\sigma^2} \int_a^b e^{\frac{-(x-(m+k\sigma^2))^2}{2\sigma^2}} \, dx \tag{C.70}$$

By substituting $x + (m + k\sigma^2)$ for x in the integrand, we obtain:

$$\int_a^b e^{kx} e^{-\frac{(x-m)^2}{2\sigma^2}} \, dx = e^{mk+\frac{k^2}{2}\sigma^2} \int_{a-m-k\sigma^2}^{b-m-k\sigma^2} e^{\frac{-x^2}{2\sigma^2}} \, dx \tag{C.71}$$

which can be integrated by equation (C.3):

$$\int_a^b e^{kx} e^{-\frac{(x-m)^2}{2\sigma^2}} \, dx \;=\; \sigma\sqrt{2\pi} e^{mk+\frac{k^2}{2}\sigma^2} \frac{1}{2} \left(\mathrm{erf}\left(\frac{b-m-k\sigma^2}{\sigma\sqrt{2}} \right) \right.$$
$$\left. - \mathrm{erf}\left(\frac{a-m-k\sigma^2}{\sigma\sqrt{2}} \right) \right) \tag{C.72}$$

Using equation (C.72), we can readily compute the expected value of $e^{\alpha X} e^{\beta Z}$, where X is normally distributed and Z is a truncated version of X. As before, the mean of $e^{\alpha X} e^{\beta Z}$ is calculated as a sum of three integrals:

$$\begin{aligned}
E(e^{\alpha X + \beta Z}) \;=\; & \frac{1}{\sigma\sqrt{2\pi}} \left[\int_{Z_{min}}^{Z_{max}} e^{(\alpha+\beta)x} e^{-\frac{(x-m)^2}{2\sigma^2}} \, dx \right. \\
& + \int_{-\infty}^{Z_{min}} e^{\beta Z_{min}} e^{\alpha x} e^{-\frac{(x-m)^2}{2\sigma^2}} \, dx \\
& \left. + \int_{Z_{max}}^{\infty} e^{\beta Z_{max}} e^{\alpha x} e^{-\frac{(x-m)^2}{2\sigma^2}} \, dx \right]
\end{aligned} \tag{C.73}$$

By simply applying equation (C.72), we have:

$$\begin{aligned}
E(e^{\alpha X + \beta Z}) \;=\; & \frac{1}{2} e^{(\alpha+\beta)m+\frac{(\alpha+\beta)^2}{2}\sigma^2} \left(\mathrm{erf}\left(\frac{Z_{max} - m - (\alpha+\beta)\sigma^2}{\sigma\sqrt{2}} \right) \right. \\
& \left. - \mathrm{erf}\left(\frac{Z_{min} - m - (\alpha+\beta)\sigma^2}{\sigma\sqrt{2}} \right) \right) \\
& + \frac{1}{2} e^{\alpha m+\frac{\alpha^2}{2}\sigma^2+\beta Z_{min}} \left(1 + \mathrm{erf}\left(\frac{Z_{min} - m - \alpha\sigma^2}{\sigma\sqrt{2}} \right) \right) \\
& + \frac{1}{2} e^{\alpha m+\frac{\alpha^2}{2}\sigma^2+\beta Z_{max}} \left(1 - \mathrm{erf}\left(\frac{Z_{max} - m - \alpha\sigma^2}{\sigma\sqrt{2}} \right) \right) \tag{C.74}
\end{aligned}$$

APPENDIX D
SOFTWARE FOR UNIT COMMITMENT

D.1   Dynamic Programming Software

The programs used for solving unit commitment in this book all run on a
UNIX platform. The principal program used is titled uci2. This program
uses enumerative dynamic programming to solve the unit commitment problem and is run by the command:

```
uci2 file
```

This command reads the problem data from file.dat and outputs the results
to file.dym. uci2 uses the mean-reverting intercept model for the price
process.

   The input file has the format shown in Table D.1. This file may be
created using any standard text editor. The first line of the file contains
the cost curve of generation. The second line gives the fixed cost while the
generator is not running. The third line has the minimum up and down times
for the generator. The fourth line gives the startup and shutdown costs. The
fifth line marks the upper and lower limits of generation. The sixth line gives
the price and load from the previous hour; this information is needed for the
price model. The seventh line contains the price process parameters. The
eighth line gives the number of days (n_d) and hours (n_h) in the time horizon.
The subsequent lines gives the expected load and standard deviation of the
load estimate for each hour in the time horizon; there are a total of n_h lines
in this section. Finally, the last line selects the discretization level of the
price state. Note that this quantity has units of the logarithm of price. The

$$a_1 \ b_1 \ c_1$$
$$c_f$$
$$t_{up} \ t_{dn}$$
$$S_1 \ T_1$$
$$P_{G1}^{min} \ P_{G1}^{max}$$
$$p_{-1} \ L_{-1}$$
$$\eta \ \bar{b} \ m \ \sigma_e$$
$$n_d \ n_h$$
$$\hat{L}_0 \ \sigma_{L0}$$
$$\hat{L}_1 \ \sigma_{L1}$$
$$\vdots$$
$$\hat{L}_{n_h-1} \ \sigma_{L(n_h-1)}$$
$$\Delta b$$

Table D.1: Format for input file (`file.dat`)

\# policies/stage
\# simulations/policy
\# policies to select (ignored)
$x_0(1)$
Sampling range of thresholds (lower upper)

Table D.2: Format for parameter file (`file.par`)

input file for the PJM numerical example solved in this book is given below.

```
2 2 18
4
3 2
4 4
5 8
13.91 26167
0.317 0.788 7.05e-5 0.1612
1 24
23830 996
22402 955
21531 925
21291 909
21374 903
22429 948
24562 1111
27274 1177
29123 1137
30396 1149
31462 1198
31989 1242
32089 1277
32252 1332
32275 1386
32145 1438
32138 1547
32024 1829
31448 1748
30982 1570
31216 1292
31092 1221
28937 1185
26167 1134
0.05
```

After uci2 is started, the program asks for the hour of the first decision, which must be a number between 0 and n_h. This number is equal to the hour for which an optimal unit commitment decision is desired; note that the price for this hour is assumed to be unknown. The program then solves the unit commitment problem for the specified data for a horizon of $n_d n_h$ hours. The load data loops continuously for each day in the horizon. If each day has different expected load data, then the entire horizon should be entered as one "day" with each hour's load profile explicitly specified.

The initial price and load data specified in the input file may be overridden by using the -b switch:

```
uci2 -b file
```

This option forces the initial price intercept value to be equal to its mean of $\bar{b}$. The -h switch provides help information on uci2:

```
uci2 -h | more
```

One error message which may appear is:

```
uci2: Discretization too small
```

If this occurs, there are two options:

- Set the discretization level to a higher value.

- If a more accurate solution is needed, then set the defined variables PSTATES and STATES to larger values and recompile the program:

  ```
  gcc uci2.c -lm -o uci2
  ```

  PSTATES is the number of price levels needed for a range of ±5 standard deviations; STATES is the total number of states at the last stage in the planning horizon.

The output file `file.dym` will look similar to the one below for the PJM example. The first lines of the file give the "on" state and "off" state thresholds of the optimal policy for each stage in the horizon. Following this section, the optimal cost-to-go and optimal decision is enumerated for each state at the second and first stages of the horizon. (Note that cost is the negative of profit.)

```
24: 0.51 0.66
23: 0.31 0.46
22: 0.21 0.46
21: 0.21 0.41
20: 0.16 0.36
19: 0.11 0.31
18: 0.11 0.31
17: 0.11 0.31
16: 0.06 0.31
15: 0.06 0.31
14: 0.11 0.31
13: 0.11 0.31
12: 0.16 0.36
11: 0.21 0.46
10: 0.36 0.56
 9: 0.51 0.76
 8: 0.81 1.01
```

```
7: 1.06 1.26
6: 1.26 1.41
5: 1.31 1.51
4: 1.26 1.51
3: 1.21 1.46
2: 1.06 1.36
1: 0.86 1.16
1: -290.07 -293.14 -296.28 -299.50 -302.79 -306.18
   -309.64 -313.20 -316.85 -320.59 -324.43 -328.37
   -332.41 -336.57 -340.84 -345.22 -349.72 -354.36
   -359.12 -364.02 -369.07 -374.27 -379.63 -385.16
   -390.87 -396.78 -402.90 -409.24 -415.83 -422.69
   -429.85 -437.33 -445.17 -453.41 -462.08 -471.22
   -480.88 -491.09 -501.91 -513.36 -525.49 -538.33
   -551.89
   -329.45 -331.59 -333.77 -336.01 -338.30 -340.64
   -343.05 -345.51 -348.04 -350.63 -353.29 -356.01
   -358.81 -361.69 -364.64 -367.67 -370.79 -374.00
   -377.32 -380.74 -384.30 -387.99 -391.85 -395.90
   -400.18 -404.71 -409.53 -414.69 -420.21 -426.13
   -432.48 -439.30 -446.61 -454.42 -462.77 -471.68
   -481.18 -491.28 -502.02 -513.43 -525.53 -538.35
   -551.90
   -370.55 -371.46 -372.38 -373.31 -374.24 -375.17
   -376.11 -377.06 -378.01 -378.97 -379.93 -380.90
   -381.88 -382.86 -383.85 -384.85 -385.85 -386.86
   -387.88 -388.91 -389.94 -390.98 -392.04 -395.90
   -400.18 -404.71 -409.53 -414.69 -420.21 -426.13
   -432.48 -439.30 -446.61 -454.42 -462.77 -471.68
   -481.18 -491.28 -502.02 -513.43 -525.53 -538.35
   -551.90
   -374.55 -375.46 -376.38 -377.31 -378.24 -379.17
   -380.11 -361.06 -382.01 -382.97 -383.93 -384.90
   -385.88 -386.86 -387.85 -388.85 -389.85 -390.86
   -391.88 -392.91 -393.94 -394.99 -396.05 -397.12
   -398.21 -399.33 -400.49 -401.70 -402.98 -404.36
   -405.88 -407.59 -409.52 -411.73 -414.29 -417.23
   -420.61 -424.45 -428.78 -433.61 -438.92 -444.70
   -450.92
   -374.55 -375.46 -376.38 -377.31 -378.24 -379.17
   -380.11 -381.06 -382.01 -382.97 -383.93 -384.90
   -385.88 -386.86 -387.85 -388.85 -389.85 -390.86
   -391.88 -392.91 -393.94 -394.99 -396.05 -397.12
   -398.21 -399.33 -400.49 -401.70 -402.98 -407.75
   -415.88 -424.31 -433.06 -442.17 -451.66 -461.58
   -471.95 -482.83 -494.25 -506.26 -518.87 -532.14
   -546.09

       1         1         1         1         1         1
       1         1         1         1         1         1
```

```
1          1          1          1          1          1
1          1          1          1          1          1
1          1          1          1          1          1
1          1          1          1          1          1
1          1          1          1          1          1
1
1          1          1          1          1          1
1          1          1          1          1          1
1          1          1          1          1          1
1          1          1          1          1          1
1          1          1          1          1          1
1          1          1          1          1          1
1          1          1          1          1          1
1
0          0          0          0          0          0
0          0          0          0          0          0
0          0          0          0          0          0
0          0          0          0          0          1
1          1          1          1          1          1
1          1          1          1          1          1
1          1          1          1          1          1
1
0          0          0          0          0          0
0          0          0          0          0          0
0          0          0          0          0          0
0          0          0          0          0          0
0          0          0          0          0          0
0          0          0          0          0          0
0          0          0          0          0          0
0
0          0          0          0          0          0
0          0          0          0          0          0
0          0          0          0          0          0
0          0          0          0          0          0
0          0          0          0          0          1
1          1          1          1          1          1
1          1          1          1          1          1
1
  0: -397.19 -402.40 -402.40 -391.28 -391.28
     1          1          1          0          0
```

D.2   Ordinal Optimization Software

The ordinal optimization program for unit commitment ucoo1 operates very similarly. The command syntax is the same as for uci2:

```
ucoo1 file
```

However, in addition to `file.dat`, which has the same format as described previously, `ucoo1` also requires a file `file.par`, which contains the parameters for the ordinal optimization. Table D.2 shows the format for this file. Note that the third entry is not relevant to the backwards iteration algorithm, and the fourth entry (current generator state) uses the same values as described in Chapter 2. As with `uci2`, the error:

`ucoo1: Out of array space`

means that either fewer policies per stage must be selected, or the program must be recompiled with a larger value of NSTATES. The parameter file for the PJM problem follows.

```
50
20
12
3
0 2
```

The output of the ordinal optimization algorithm is in `file.oo`. The output for the PJM example is shown below. The first line gives the average simulated cost for the "off" decision and its variance, followed by the simulated cost of the "on" decision and its variance. The optimal decision is in parentheses. Next comes the mean simulated difference between the two decisions, the variance of this difference, and finally, the confidence level that the average difference is greater than (or less than) zero. The variances in this line are all for one simulation; the variance of the means is obtained by dividing by the number of simulations. The remaining lines in the output file give the selected policy.

```
-394.30 131665.80 -408.09 133828.44 (1) 13.79 275.24 99.98%
   1 -- 0.82 1.03
   2 -- 1.08 1.99
   3 -- 1.43 1.57
   4 -- 1.07 1.81
   5 -- 1.46 1.70
   6 -- 1.56 2.00
   7 -- 1.02 1.78
   8 -- 0.69 1.05
   9 -- 0.56 0.91
  10 -- 0.22 0.39
  11 -- 0.08 0.66
  12 -- 0.07 0.17
  13 -- 0.12 0.23
  14 -- 0.26 0.49
  15 -- 0.12 0.48
```

```
16 -- 0.07 0.07
17 -- 0.02 0.35
18 -- 0.20 0.26
19 -- 0.25 0.27
20 -- 0.05 0.33
21 -- 0.26 0.34
22 -- 0.08 0.50
23 -- 0.37 0.46
24 -- 0.55 0.74
```

D.3   Reserve and Congestion DP Software

Finally, there are two developmental program versions that include reserve and congestion information. uci2rs includes reserve price data from file.rdt, in the format of Table D.3. The first line gives respectively the probabilities of reserve calls and a generator failure; the second line gives the parameters for the hypothetical reserve price model. uci2con uses congestion limit probabilities specified in the file file.cdt, with the format of Table D.4. The variable P_{Clim} is treated as an exogenous random input independent of the price, which takes on discrete values with given probabilities. The first line of file.cdt is the number of possible values of P_{Clim}; each value is enumerated along with its probability on the remaining lines. Examples of these files are shown below. The reserve data file for the PJM example is:

```
0.005 0.0001
0.7 0.25
```

The congestion data file is:

```
3
1000 0.8
7 0.1
5 0.1
```

$$r\ f$$
$$K_R\ \sigma_R$$

Table D.3: Format for reserve data file (`file.rdt`)

$$n_{Clim}$$
$$V_1\ \mathrm{Prob}(P_{Clim} = V_1)$$
$$\vdots$$
$$V_{n_{Clim}}\ \mathrm{Prob}(P_{Clim} = V_{n_{Clim}})$$

Table D.4: Format for congestion data file (`file.cdt`)

D.4   Source Code for uci2.c

```c
/*  Name:  Eric Allen
    Date:  12-2-97
    Description: Program to implement the dynamic programming algorithm for
    unit commitment for an individual power producer, using price process
    model with load forecasts and exponential algorithm.  Log of price is
    modeled as a mean-reverting intercept.
*/

#include<stdio.h>
#include<math.h>
#include<stdlib.h>
#include<string.h>

#include"ucerr.h"
/* #include"uc1.h" */

#define EPSILON 0.0000001
#define PI 3.14159265358979

#define MAXP 96
#define TSTEPS 10
#define STATES 10000
#define PSTATES 500
#define ALT 3

double erf (double z) {

  double term, temp, fact, sum;
  int i;

  if (z <= -3.5)
    return (-1);
  if (z >= 3.5)
    return (1);

  i = 1;
  fact = 1;
  sum = z;

  do {
    fact = fact*i;
    temp = (2.0*i + 1);
    term = pow(z, temp)/ temp / fact;
    if (2 * (i/2) != i++)
      sum -= term;
    else
      sum += term;
  } while ((fabs(term) > EPSILON) && (i < 100));
  return (2*sum/sqrt(PI));
}

/* Calculates the next state from the current state and control option.
   There are minup + mindn possible states.  States 0, 1, ..., minup-1
   correspond to the generator having been up 1, 2, ..., minup periods, while
   states minup, minup+1, ..., minup+mindn-1 correspond to the generator
```

```
  having been down 1, 2, ..., mindn periods. */

int nextst(int currst, int opt, int mup, int mdn) {

  int n;

  if (currst < mup)
    if (opt == 0)
      n = mup;
    else if (currst < mup - 1)
      n = currst + 1;
    else
      n = currst;
  else
    if (opt)
      n = 0;
    else if (currst < mup + mdn - 1)
      n = currst + 1;
    else
      n = currst;
  return(n);
}

main(int argc, char *argv[])
{
  double ca, cb, cc;      /* cost curve coefficients */
  double coff;            /* expected cost per stage while off */
  int minup, mindn;       /* Minimum on/off times */
  double cup, cdn;        /* Startup/shutdown costs */
  double pmin, pmax;      /* Minimum/maximum generation levels */

  double exld[MAXP], stdld[MAXP];  /* Mean/variance of load forecast */

  double ex, stdev;       /* Mean/variance of price */
  double expr;            /* Expected price */
  double extr, vtr, sq2;  /* Mean/variance of truncated variable; sqrt(2) */
  double exprod;          /* Expected product of price and marginal cost */
  double pmcmin, pmcmax;  /* Marginal cost limits */
  double zl, zu;          /* Z- and Z+ (normalized limits) */

  double Jk[2][STATES];         /* cost-to-go function */
  int uk[2][STATES];            /* optimal control */
  double probmat[PSTATES];      /* Transition probabilities */

  double *Jc, *J1, *Jtmp;       /* current and next state cost-to-go */
  int *uc, *u1, *utmp;          /* current and next state control */

  int index;            /* index array to find state no. */
  int state;            /* state counter */
  int hour;             /* First hour in time horizon */

  double ecost[STATES];         /* Expected cost for one stage */
  int lu;                       /* Minimizing control */
  double lcost;                 /* Cost of minimizing control */

  double b0;                    /* Starting intercept */
  double eta, bbar, m, sigma;   /* Price process parameters */
```

```
    double disc;                    /* Discretization level */
    int nstdsc;                     /* No. of discretized states */
    int nsttot;                     /* Total number of price states */
    int iref;                       /* Reference into price state array */
    int stlim;                      /* Price state limit for given stage */

    int i, j, k, k1, nd, np, nst, hlen, nalt, nopt, baseopt, curropt, tmp, bit;
    int next0, next1;               /* Next state for 0/1 choice of control */
    double tmpd, tmpd1, tmpd2, tmpd3, tmpd4;
    char check;

    int prgopt = 0;                 /* Program options selected */
    int dot;                        /* Length of filename (without extender) */
    char *filename;
    FILE *fp;

    /* nd:      number of days in horizon
       np:      number of periods per day
       nst:     number of states
       nalt:    number of control alternatives
       nopt:    number of control options available for a given state
       baseopt: control alternative with no generator status changed
       curropt: current control alternative being examined
       hlen:    length of horizon
       bit:     bit mask to calculate control alternative number
       stlim:   limit of number of price states at current stage
    */

/* Process command line arguments */

  if ((argc < 2) || (argc > 3))
    error(argv[0], argv[0], 1);   /* Syntax error */
  if (argc == 3)
    if (strlen(argv[1]) != 2)
      error(argv[0], argv[0], 1); /* Syntax error */
    else if (argv[1][0] != '-')
      error(argv[0], argv[0], 1); /* Syntax error */
    else {
      switch(argv[1][1]) {
      case 'b': prgopt = 1; break;
      case 'h': help(argv[0]);
      case 'o': prgopt = 2; break;
      default: error(argv[0], argv[0], 1); /* Syntax error */
      }
      tmp = 2;
    }
  else if (argv[1][0] == '-')
    if (strlen(argv[1]) != 2)
      error(argv[0], argv[0], 1); /* Syntax error */
    else if (argv [1][1] == 'h')
      help(argv[0]);
    else
      error(argv[0], argv[0], 1); /* Syntax error */
  else
    tmp = 1;
  dot = strlen(argv[tmp]);
```

```
  filename = malloc((dot + 5)*sizeof(char));
  if (filename == NULL)
    error(argv[0], argv[0], 4); /* Out of memory */
  strcpy(filename, argv[tmp]);

/* Load input data */

  strcat(filename, ".dat");
  fp = fopen(filename,"r");
  if (fp == NULL)
    error(argv[0], filename, 2);   /* File not found */
  fscanf(fp, "%lf %lf %lf", &ca, &cb, &cc);
  fscanf(fp, "%lf", &coff);
  fscanf(fp, "%d %d", &minup, &mindn);
  fscanf(fp, "%lf %lf", &cup, &cdn);
  fscanf(fp, "%lf %lf", &pmin, &pmax);
  fscanf(fp, "%lf %lf", &tmpd, &tmpd1);
  fscanf(fp, "%lf %lf %lf %lf", &eta, &bbar, &m, &sigma);
  b0 = log(tmpd) - m*tmpd1;
  if (prgopt == 1)
    b0 = bbar;
  fscanf(fp, "%d %d", &nd, &np);
  for (i = 0; i < np; i++)
    fscanf(fp, "%lf %lf", exld+i, stdld+i);
  fscanf(fp, "%lf", &disc);
  fclose(fp);

  printf ("Enter hour of first decision (0-%d):",np);
  scanf ("%d", &hour);

/* Initialize dynamic programming */

    hlen = nd*np;
    tmpd = stdld[0];
    for (i = 1; i < np; i++)
      if (tmpd < stdld[i])
        tmpd = stdld[i];
    stdev = sqrt(sigma*sigma + m*m*tmpd*tmpd);
    nstdsc = ceil(3.5*sqrt(2)*stdev/disc);
    nsttot = 2*nstdsc*(hlen+1) + 1;
    iref = nstdsc*(hlen + 1);
    nst = minup + mindn;
    if (2*nstdsc+1 > PSTATES)
      error(argv[0], "1", 6); /* Not enough states for discretization level */
    if (nst*nsttot > STATES)
      error(argv[0], "1", 6); /* Not enough states for discretization level */
    Jc = Jk[0]; J1 = Jk[1];
    uc = uk[0]; u1 = uk[1];
    for (i = 0; i < nst*nsttot; i++)
      J1[i] = 0;
    filename[dot] = '\0';
    strcat(filename, ".dym");
    fp = fopen(filename,"w");
    if (fp == NULL)
      error(argv[0], argv[2], 3);   /* Unable to write to file */
    printf("Performing dynamic programming . . .\n");
    sq2 = sqrt(2);
```

```
      pmcmin = 2*ca*pmin + cb;
      pmcmax = 2*ca*pmax + cb;
      stlim = nstdsc*hlen;

/* Dynamic Programming */

      for (i = hlen; i >= 0; i--) {
        for (j = -stlim; j <= stlim; j++) {
          index = (i + hour) % np;
          ex = exp(-eta)*(b0 + j*disc - bbar) + bbar + m*exld[index];
          stdev = sqrt(sigma*sigma + m*m*stdld[index]*stdld[index]);
          expr = exp(ex + 0.5*stdev*stdev);
          zl = (log(pmcmin) - ex)/stdev;
          zu = (log(pmcmax) - ex)/stdev;
          extr = pmcmin*(0.5 + 0.5*erf(zl/sq2))
            + pmcmax*(0.5 - 0.5*erf(zu/sq2));
          extr = extr + 0.5*expr*(erf((zu - stdev)/sq2)
                                  - erf((zl - stdev)/sq2));
          vtr = 0.5*exp(2*ex+2*stdev*stdev)*(erf((zu-2*stdev)/sq2)
                                  - erf((zl-2*stdev)/sq2));
          vtr = vtr + pmcmin*pmcmin*(0.5 + 0.5*erf(zl/sq2));
          vtr = vtr + pmcmax*pmcmax*(0.5 - 0.5*erf(zu/sq2)) -extr*extr;
          exprod = 0.5*exp(2*ex+2*stdev*stdev)*(erf((zu-2*stdev)/sq2)
                                  - erf((zl-2*stdev)/sq2));
          exprod = exprod
            + pmcmin*exp(ex+0.5*stdev*stdev)*(0.5 + 0.5*erf((zl-stdev)/sq2));
          exprod = exprod
            + pmcmax*exp(ex+0.5*stdev*stdev)*(0.5 - 0.5*erf((zu-stdev)/sq2));
          ecost[j+iref] = (extr*extr + vtr - cb*cb + 2*cb*expr
                          - 2*exprod)/(4*ca) + cc;
        }
        for (j = 0; j < nst; j++) {
          next1 = nextst(j,1,minup,mindn);
          next0 = nextst(j,0,minup,mindn);
          for (k = -stlim; k <= stlim; k++) {
            tmpd = stdev*sq2;
            ex = exp(-eta)*(b0 + k*disc - bbar) + bbar;
            tmpd1 = erf((b0 + (k - 0.5)*disc - ex)/tmpd);
            tmpd2 = erf((b0 + (k + 0.5)*disc - ex)/tmpd);
            probmat[nstdsc] = (tmpd2 - tmpd1)/2;
            for (k1 = 1; k1 <= nstdsc; k1++) {
              tmpd3 = erf((b0 + (k-k1-0.5)*disc - ex)/tmpd);
              tmpd4 = erf((b0 + (k+k1+0.5)*disc - ex)/tmpd);
              probmat[nstdsc+k1] = (tmpd4 - tmpd2)/2;
              probmat[nstdsc-k1] = (tmpd1 - tmpd3)/2;
              tmpd1 = tmpd3;
              tmpd2 = tmpd4;
            }
            if (j < minup) {
              lu = 1;
              lcost = ecost[k+iref];
              for (k1 = -nstdsc; k1 <= nstdsc; k1++) {
                index = nsttot*next1 + k1 + k + iref;
                lcost += probmat[k1+nstdsc]*J1[index];
              }
              if (j == minup - 1) {
                tmpd = coff + cdn;
```

```
        for (k1 = -nstdsc; k1 <= nstdsc; k1++) {
          index = nsttot*next0 + k1 + k + iref;
          tmpd += probmat[k1+nstdsc]*J1[index];
        }
        if (tmpd < lcost) {
          lu = 0;
          lcost = tmpd;
        }
      }
    }
    else {
      lu = 0;
      lcost = coff;
      for (k1 = -nstdsc; k1 <= nstdsc; k1++) {
        index = nsttot*next0 + k1 + k + iref;
        lcost += probmat[k1+nstdsc]*J1[index];
      }
      if (j == nst - 1) {
        tmpd = ecost[k+iref] + cup;
        for (k1 = -nstdsc; k1 <= nstdsc; k1++) {
          index = nsttot*next1 + k1 + k + iref;
          tmpd += probmat[k1+nstdsc]*J1[index];
        }
        if (tmpd < lcost) {
          lu = 1;
          lcost = tmpd;
        }
      }
    }
    index = nsttot*j + k + iref;
    Jc[index] = lcost;
    uc[index] = lu;
  }
}
if (i >= 1) {
  index = nsttot*(minup - 1) + iref;
  fprintf (fp, "%3d: ", i);
  if (uc[index-stlim] == 1)
    fprintf (fp, "%3.2f # ", b0-(stlim+0.5)*disc);
  else if (uc[index+stlim] == 0)
    fprintf (fp, "%3.2f # ", b0+(stlim+0.5)*disc);
  else {
    k = 0;
    k1 = stlim/2;
    while ((uc[index+k] == 1) || (uc[index+k+1] == 0)) {
      if (uc[index+k] == 0)
        k += k1;
      else
        k -= k1;
      k1 = k1/2;
      if (k1 == 0)
        k1 = 1;
    }
    fprintf (fp, "%3.2f ", b0+(k+0.5)*disc);
  }
  index = nsttot*(minup + mindn - 1) + iref;
  if (uc[index-stlim] == 1)
```

```
          fprintf (fp, "%3.2f # ", b0-(stlim+0.5)*disc);
        else if (uc[index+stlim] == 0)
          fprintf (fp, "%3.2f # ", b0+(stlim+0.5)*disc);
        else {
          k = 0;
          k1 = stlim/2;
          while ((uc[index+k] == 1) || (uc[index+k+1] == 0)) {
            if (uc[index+k] == 0)
              k += k1;
            else
              k -= k1;
            k1 = k1/2;
            if (k1 == 0)
              k1 = 1;
          }
          fprintf (fp, "%3.2f ", b0+(k+0.5)*disc);
        }
        fprintf (fp, "\n");
    }
    stlim -= nstdsc;
    printf("%3d ... ", i);
    if (i == 1) {
      fprintf(fp,"%3d: ",i);
      for (j = 0; j < nst; j++) {
        for (k = -nstdsc; k <= nstdsc; k++) {
          index = j*nsttot + k + iref;
          fprintf(fp, "%7.2f ", Jc[index]);
          if ((k+nstdsc) % 6 == 5)
            fprintf(fp, "\n        ");
        }
        fprintf(fp,"\n        ");
      }
      fprintf(fp, "\n        ");
      for (j = 0; j < nst; j++) {
        for (k = -nstdsc; k <= nstdsc; k++) {
          index = j*nsttot + k + iref;
          fprintf(fp, "%10d ", uc[index]);
          if ((k+nstdsc) % 6 == 5)
            fprintf(fp, "\n        ");
        }
        fprintf(fp,"\n        ");
      }
      printf("\n(%d %d) ... ",stlim, abs(-nstdsc));
    }
    Jtmp = Jc; Jc = J1; J1 = Jtmp;
    utmp = uc; uc = uc; u1 = utmp;
}
fprintf(fp,"%3d: ",0);
for (j = 0; j < nst; j++) {
  fprintf(fp, "%7.2f ", J1[j*nsttot + iref]);
  if (j % 6 == 5)
    fprintf(fp, "\n        ");
}
fprintf(fp,"\n        ");
for (j = 0; j < nst; j++) {
  fprintf(fp, "%10d ", u1[j*nsttot + iref]);
  if (j % 6 == 5)
```

```
        fprintf(fp, "\n      ");
    }
    fprintf(fp,"\n      ");
    printf("\n");
    fclose(fp);
}
```

D.5   Source Code for ucoo1.c

```c
/*  Name:  Eric Allen
    Date:  3-25-98
    Description: Program to implement the dynamic programming algorithm for
    unit commitment for an individual power producer, using price process
    model with load forecasts.  Log of price is
    modeled as a mean-reverting intercept.  This program has a simulation
    engine and uses ordinal optimization.
*/

#include<stdio.h>
#include<math.h>
#include<stdlib.h>
#include<string.h>

#include"ucerroo.h"
/* #include"uc1.h" */

#define EPSILON 0.0000001
#define PI 3.14159265358979

#define MAXP 96
#define NSAMPLE 100

/* Return a pseudo-random number between 0 and 1 */

double urnd(void) {

  int u;

  u = rand();
  return (((double)u)/RAND_MAX);
}

/* Return a pseudo-random normal variable with mean 0 and variance 1
   using the polar method */

double normrnd(void) {

  static int state = 0;
  static double n2 = 0;
  double u1, u2, v1, v2, s;

  if (state) {
    state = 0;
    return(n2);
  }
  else {
    do {
      u1 = urnd(); u2 = urnd();
      v1 = 2*u1 - 1;
      v2 = 2*u2 - 1;
      s = v1*v1 + v2*v2;
    } while ((s > 1) || (s == 0));
    u1 = sqrt(-2*log(s)/s);
    n2 = u1*v2;
```

```
      state = 1;
      return (u1*v1);
  }
}

double erf (double z) {

  double term, temp, fact, sum;
  int i;

  if (z <= -3.5)
    return (-1);
  if (z >= 3.5)
    return (1);

  i = 1;
  fact = 1;
  sum = z;

  do {
    fact = fact*i;
    temp = (2.0*i + 1);
    term = pow(z, temp)/ temp / fact;
    if (2 * (i/2) != i++)
      sum -= term;
    else
      sum += term;
  } while ((fabs(term) > EPSILON) && (i < 100));
  return (2*sum/sqrt(PI));
}

/* Calculates the next state from the current state and control option.
   There are minup + mindn possible states.  States 1, 2, ..., minup
   correspond to the generator having been up 1, 2, ..., minup periods, while
   states -1, -2, ..., -mindn correspond to the generator
   having been down 1, 2, ..., mindn periods. */

int nextst(int currst, int opt, int mup, int mdn) {

  int n;

  if (currst > 0)
    if (opt == 0)
      n = -1;
    else if (currst < mup)
      n = currst + 1;
    else
      n = currst;
  else
    if (opt)
      n = 1;
    else if (currst > -mdn)
      n = currst - 1;
    else
      n = currst;
  return(n);
}
```

```
main(int argc, char *argv[])
{
  double ca, cb, cc;      /* cost curve coefficients */
  double coff;            /* expected cost per stage while off */
  int minup, mindn;       /* Minimum on/off times */
  double cup, cdn;        /* Startup/shutdown costs */
  double pmin, pmax;      /* Minimum/maximum generation levels */

  double exld[MAXP], stdld[MAXP];  /* Mean/variance of load forecast */

  double sq2;             /* sqrt(2) */

  double polup[NSAMPLE];  /* Thresholds of policy while up */
  double poldn[NSAMPLE];  /* Thresholds of policy while down */

  double thrup[MAXP], thrdn[MAXP];  /* Up/down thresholds of current policy */

  double horse[NSAMPLE];  /* Simulated cost of policy */
  double simsd[NSAMPLE];  /* Std. dev. of simulated cost */

  int state0;        /* Initial state of simulation */
  int dec0;          /* Initial decision in simulation */

  int npol;          /* Number of policies in sample (N) */
  int nsim;          /* Number of simulations per policy in sample */
  int sel;           /* No. of selected policies from sample (s) */
  double smplo, smphi;  /* Lower/upper bounds on policy sampling */

  int index;         /* index array to find state no. */
  int xstate;        /* state counter */
  int hour;          /* First hour in time horizon */
  double tcost;      /* Cost in simulation */
  double pg;         /* Generation level */
  double pr, b;      /* Price, intercept in simulation */
  double bsim, prsim;  /* Starting intercept/price in simulation */

  double ss[NSAMPLE], ssq[NSAMPLE];  /* Sum, sum of squares */
  double rdata[MAXP][2];             /* Random variables for simulation */

  double b0;              /* Starting intercept */
  double eta, bbar, m, sigma;  /* Price process parameters */
  double eeta;            /* Exponential of eta */

  int i, j, k, k1, nd, np, nst, hlen, nalt, nopt, baseopt, curropt, tmp, bit;
  int next0, next1;       /* Next state for 0/1 choice of control */
  double tmpd, tmpd1, tmpd2, tmpd3, tmpd4, diff;
  char check;

  int prgopt = 0;         /* Program options selected */
  int dot;                /* Length of filename (without extender) */
  char *filename;
  FILE *fp;

  /* nd:    number of days in horizon
     np:    number of periods per day
     nst:   number of states
```

```
    nalt:    number of control alternatives
    nopt:    number of control options available for a given state
    baseopt: control alternative with no generator status changed
    curropt: current control alternative being examined
    hlen:    length of horizon
    bit:     bit mask to calculate control alternative number
    stlim:   limit of number of price states at current stage
  */

/* Process command line arguments */

  if ((argc < 2) || (argc > 3))
    error(argv[0], argv[0], 1);   /* Syntax error */
  if (argc == 3)
    if (strlen(argv[1]) != 2)
      error(argv[0], argv[0], 1); /* Syntax error */
    else if (argv[1][0] != '-')
      error(argv[0], argv[0], 1); /* Syntax error */
    else {
      switch(argv[1][1]) {
      case 'b': prgopt = 1; break;
      case 'h': help(argv[0]);
      case 'o': prgopt = 2; break;
      default: error(argv[0], argv[0], 1); /* Syntax error */
      }
      tmp = 2;
    }
  else if (argv[1][0] == '-')
    if (strlen(argv[1]) != 2)
      error(argv[0], argv[0], 1); /* Syntax error */
    else if (argv [1][1] == 'h')
      help(argv[0]);
    else
      error(argv[0], argv[0], 1); /* Syntax error */
  else
    tmp = 1;
  dot = strlen(argv[tmp]);
  filename = malloc((dot + 5)*sizeof(char));
  if (filename == NULL)
    error(argv[0], argv[0], 4); /* Out of memory */
  strcpy(filename, argv[tmp]);

/* Load input data */

  strcat(filename, ".dat");
  fp = fopen(filename,"r");
  if (fp == NULL)
    error(argv[0], filename, 2);  /* File not found */
  fscanf(fp, "%lf %lf %lf", &ca, &cb, &cc);
  fscanf(fp, "%lf", &coff);
  fscanf(fp, "%d %d", &minup, &mindn);
  fscanf(fp, "%lf %lf", &cup, &cdn);
  fscanf(fp, "%lf %lf", &pmin, &pmax);
  fscanf(fp, "%lf %lf", &tmpd, &tmpd1);
  fscanf(fp, "%lf %lf %lf %lf", &eta, &bbar, &m, &sigma);
  b0 = log(tmpd) - m*tmpd1;
  if (prgopt == 1)
```

```
   b0 = bbar;
 eeta = exp(-eta);
 fscanf(fp, "%d %d", &nd, &np);
 for (i = 0; i < np; i++)
   fscanf(fp, "%lf %lf", exld+i, stdld+i);
 fscanf(fp, "%lf", &tmpd);  /* Read and discard discretization level */
 fclose(fp);

 filename[dot] = '\0';
 strcat(filename, ".par");
 fp = fopen(filename,"r");
 if (fp == NULL)
   error(argv[0], filename, 2);  /* File not found */
 fscanf(fp, "%d", &npol);
 fscanf(fp, "%d", &nsim);
 fscanf(fp, "%d", &sel);
 fscanf(fp, "%d", &state0);
 fscanf(fp, "%lf %lf", &smplo, &smphi);
 fclose(fp);
 printf ("Enter hour of first decision (0-%d):",np);
 scanf ("%d", &hour);

/* Initialize ordinal optimization */

 hlen = nd*np;
 if (hlen > MAXP)
   error(argv[0], "1", 6); /* Not enough states for horizon length */
 if (2*npol > NSAMPLE)
   error(argv[0], "1", 6); /* Not enough space for policy sample */
 filename[dot] = '\0';
 strcat(filename, ".oo");
 fp = fopen(filename,"w");
 if (fp == NULL)
   error(argv[0], argv[2], 3);    /* Unable to write to file */
 printf("Performing ordinal optimization . . .\n");
 sq2 = sqrt(2);

/* Ordinal Optimization */

    for (i = hlen - 1; i >= 0; i--) {
      for (j = 0; j < npol; j++) {
        ss[j] = 0; ssq[j] = 0;
        tmpd = (smphi - smplo)*urnd() + smplo;
        tmpd1 = (smphi - smplo)*urnd() + smplo;
        if (tmpd < tmpd1) {
          polup[j] = tmpd;
          poldn[j] = tmpd1;
        }
        else {
          poldn[j] = tmpd;
          polup[j] = tmpd1;
        }
      }
      for (k1 = 0; k1 < 2*nsim; k1++) {
        bsim = exp(-(i+1)*eta)*(b0 - bbar) + bbar
          + sigma*sqrt((1 - exp(-2*(i+1)*eta))/(1 - exp(-2*eta)))*normrnd();
        b = bsim;
```

```
      for (j = i; j < hlen; j++) {
        index = (j + 1 + hour) % np;
        b = (1 - eeta)*bbar + eeta*b + sigma*normrnd();
        pr = exp(b + m*(exld[index] + stdld[index]*normrnd()));
        rdata[j][0] = b;
        rdata[j][1] = pr;
      }
      for (k = 0; k < npol; k++) {
        thrup[i] = polup[k];
        thrdn[i] = poldn[k];
        xstate = minup;
        if (k1 >= nsim)
          xstate = -mindn;

/* Simulation engine */

        b = bsim;
        tcost = 0;
        for (j = i; j < hlen; j++) {
          if (xstate > 0) {
            if ((b < thrup[j]) && (xstate >= minup)) {
              tcost += cdn;
              xstate = -1;
            }
            else if (xstate < minup)
              xstate++;
          }
          else {
            if ((b > thrdn[j]) && (xstate <= -mindn)) {
              tcost += cup;
              xstate = 1;
            }
            else if (xstate > -mindn)
              xstate--;
          }
          b = rdata[j][0];
          pr = rdata[j][1];
          if (xstate > 0) {
            pg = (pr - cb)/(2*ca);
            if (pg < pmin)
              pg = pmin;
            if (pg > pmax)
              pg = pmax;
            tcost += ca*pg*pg + cb*pg + cc - pr*pg;
          }
          else
            tcost += coff;
        }
        ss[k] += tcost;
        ssq[k] += tcost*tcost;
      }
    }
    for (k = 0; k < npol; k++) {
      horse[k] = ss[k]/(2*nsim);
      simsd[k] = (ssq[k] - ss[k]*ss[k]/(2*nsim))/(2*nsim - 1);
    }
    tmp = 0;
```

```
      tmpd = horse[0];
      for (j = 1; j < npol; j++)
        if (horse[j] < tmpd) {
           tmp = j;
           tmpd = horse[j];
        }
      thrup[i] = polup[tmp];
      thrdn[i] = poldn[tmp];
   }
   for (j = 0; j <= 2; j++) {
      ss[j] = 0; ssq[j] = 0;
   }
   for (k1 = 0; k1 < nsim; k1++) {
      b = b0;
      index = hour % np;
      bsim = (1 - eeta)*bbar + eeta*b + sigma*normrnd();
      prsim = exp(b + m*(exld[index] + stdld[index]*normrnd()));
      b = bsim;
      for (j = 0; j < hlen; j++) {
         index = (j + 1 + hour) % np;
         b = (1 - eeta)*bbar + eeta*b + sigma*normrnd();
         pr = exp(b + m*(exld[index] + stdld[index]*normrnd()));
         rdata[j][0] = b;
         rdata[j][1] = pr;
      }
      for (dec0 = 0; dec0 <= 1; dec0++) {
         xstate = state0;

/* Simulation engine at initial stage */

         tcost = 0;
         if (xstate > 0) {
            if ((dec0 == 0) && (xstate >= minup)) {
               tcost += cdn;
               xstate = -1;
            }
            else if (xstate < minup)
               xstate++;
         }
         else {
            if ((dec0 == 1) && (xstate <= -mindn)) {
               tcost += cup;
               xstate = 1;
            }
            else if (xstate > -mindn)
               xstate--;
         }
         b = bsim;
         pr = prsim;
         if (xstate > 0) {
            pg = (pr - cb)/(2*ca);
            if (pg < pmin)
               pg = pmin;
            if (pg > pmax)
               pg = pmax;
            tcost += ca*pg*pg + cb*pg + cc - pr*pg;
         }
```

```
      else
        tcost += coff;
      for (j = 0; j < hlen; j++) {
        if (xstate > 0) {
          if ((b < thrup[j]) && (xstate >= minup)) {
            tcost += cdn;
            xstate = -1;
          }
          else if (xstate < minup)
            xstate++;
        }
        else {
          if ((b > thrdn[j]) && (xstate <= -mindn)) {
            tcost += cup;
            xstate = 1;
          }
          else if (xstate > -mindn)
            xstate--;
        }
        b = rdata[j][0];
        pr = rdata[j][1];
        if (xstate > 0) {
          pg = (pr - cb)/(2*ca);
          if (pg < pmin)
            pg = pmin;
          if (pg > pmax)
            pg = pmax;
          tcost += ca*pg*pg + cb*pg + cc - pr*pg;
        }
        else
          tcost += coff;
      }
      ss[dec0] += tcost;
      ssq[dec0] += tcost*tcost;
      if (dec0)
        diff -= tcost;
      else
        diff = tcost;
    }
    ss[2] += diff;
    ssq[2] += diff*diff;
  }
  for (dec0 = 0; dec0 <= 1; dec0++) {
    horse[dec0] = ss[dec0]/nsim;
    if (nsim > 1)
      simsd[dec0] = (ssq[dec0] - ss[dec0]*ss[dec0]/nsim)/(nsim - 1);
  }
  fprintf(fp, "%3.2f %3.2f ", horse[0], simsd[0]);
  fprintf(fp, "%3.2f %3.2f ", horse[1], simsd[1]);
  if (horse[0] < horse[1])
    fprintf(fp, "(0) ");
  else
    fprintf(fp, "(1) ");
  tmpd3 = ss[2]/nsim;
  if (nsim > 1)
    tmpd4 = (ssq[2] - ss[2]*ss[2]/nsim)/(nsim - 1);
  tmpd = sqrt(tmpd4/nsim);
```

```
    tmpd1 = 0.5 + 0.5*erf(abs(tmpd3)/(tmpd*sq2));
    fprintf(fp, "%3.2f %3.2f %3.2f%%\n", tmpd3, tmpd4, tmpd1*100);
    for (j = 0; j < hlen; j++)
      fprintf(fp, "  %d -- %3.2f %3.2f\n", j+1,
                thrup[j], thrdn[j]);
    fclose(fp);
}
```

D.6 Source Code for `uci2rs.c`

```c
/*  Name:  Eric Allen
    Date:  3-24-98
    Description: Program to implement the dynamic programming algorithm for
    unit commitment for an individual power producer, using price process
    model with load forecasts and exponential algorithm.  Log of price is
    modeled as a mean-reverting intercept.  Includes reserve market data.
*/

#include<stdio.h>
#include<math.h>
#include<stdlib.h>
#include<string.h>

#include"ucerr.h"
/* #include"uc1.h" */

#define EPSILON 0.0000001
#define PI 3.14159265358979

#define MAXP 96
#define TSTEPS 10
#define STATES 10000
#define PSTATES 500
#define ALT 3

double erf (double z) {

  double term, temp, fact, sum;
  int i;

  if (z <= -3.5)
    return (-1);
  if (z >= 3.5)
    return (1);

  i = 1;
  fact = 1;
  sum = z;

  do {
    fact = fact*i;
    temp = (2.0*i + 1);
    term = pow(z, temp)/ temp / fact;
    if (2 * (i/2) != i++)
      sum -= term;
    else
      sum += term;
  } while ((fabs(term) > EPSILON) && (i < 100));
  return (2*sum/sqrt(PI));
}

/* Calculates the next state from the current state and control option.
   There are minup + mindn possible states.  States 0, 1, ..., minup-1
   correspond to the generator having been up 1, 2, ..., minup periods, while
   states minup, minup+1, ..., minup+mindn-1 correspond to the generator
   having been down 1, 2, ..., mindn periods. */
```

```
int nextst(int currst, int opt, int mup, int mdn) {

  int n;

  if (currst < mup)
    if (opt == 0)
      n = mup;
    else if (currst < mup - 1)
      n = currst + 1;
    else
      n = currst;
  else
    if (opt)
      n = 0;
    else if (currst < mup + mdn - 1)
      n = currst + 1;
    else
      n = currst;
  return(n);
}

main(int argc, char *argv[])
{
  double ca, cb, cc;      /* cost curve coefficients */
  double coff;            /* expected cost per stage while off */
  int minup, mindn;       /* Minimum on/off times */
  double cup, cdn;        /* Startup/shutdown costs */
  double pmin, pmax;      /* Minimum/maximum generation levels */

  double exld[MAXP], stdld[MAXP];  /* Mean/variance of load forecast */

  double ex, stdev;       /* Mean/variance of price */
  double expr;            /* Expected price */
  double extr, vtr, sq2;  /* Mean/variance of truncated variable; sqrt(2) */
  double exprod;          /* Expected product of price and marginal cost */
  double pmcmin, pmcmax;  /* Marginal cost limits */
  double zl, zu;          /* Z- and Z+ (normalized limits) */

  double Jk[2][STATES];           /* cost-to-go function */
  int uk[2][STATES];              /* optimal control */
  double probmat[PSTATES];        /* Transition probabilities */

  double *Jc, *J1, *Jtmp;         /* current and next state cost-to-go */
  int *uc, *u1, *utmp;            /* current and next state control */

  int index;          /* index array to find state no. */
  int state;          /* state counter */
  int hour;           /* First hour in time horizon */

  double ecost[STATES];           /* Expected cost for one stage */
  int lu;                         /* Minimizing control */
  double lcost;                   /* Cost of minimizing control */

  double b0;                      /* Starting intercept */
  double eta, bbar, m, sigma;     /* Price process parameters */
```

```
double disc;                 /* Discretization level */
int nstdsc;                  /* No. of discretized states */
int nsttot;                  /* Total number of price states */
int iref;                    /* Reference into price state array */
int stlim;                   /* Price state limit for given stage */

double prr;                  /* Probability of reserve call */
double prf;                  /* Probability of generator failure */
double Kr, sigr;             /* Reserve price constant, std. dev. */
double exexper;              /* Expectation of e^(e_R) */

int i, j, k, k1, nd, np, nst, hlen, nalt, nopt, baseopt, curropt, tmp, bit;
int next0, next1;            /* Next state for 0/1 choice of control */
double tmpd, tmpd1, tmpd2, tmpd3, tmpd4;
char check;

int prgopt = 0;              /* Program options selected */
int dot;                     /* Length of filename (without extender) */
char *filename;
FILE *fp;

/* nd:     number of days in horizon
   np:     number of periods per day
   nst:    number of states
   nalt:   number of control alternatives
   nopt:   number of control options available for a given state
   baseopt: control alternative with no generator status changed
   curropt: current control alternative being examined
   hlen:   length of horizon
   bit:    bit mask to calculate control alternative number
   stlim:  limit of number of price states at current stage
*/

/* Process command line arguments */

  if ((argc < 2) || (argc > 3))
    error(argv[0], argv[0], 1);   /* Syntax error */
  if (argc == 3)
    if (strlen(argv[1]) != 2)
      error(argv[0], argv[0], 1); /* Syntax error */
    else if (argv[1][0] != '-')
      error(argv[0], argv[0], 1); /* Syntax error */
    else {
      switch(argv[1][1]) {
      case 'b': prgopt = 1; break;
      case 'h': help(argv[0]);
      case 'o': prgopt = 2; break;
      default: error(argv[0], argv[0], 1); /* Syntax error */
      }
      tmp = 2;
    }
  else if (argv[1][0] == '-')
    if (strlen(argv[1]) != 2)
      error(argv[0], argv[0], 1); /* Syntax error */
    else if (argv [1][1] == 'h')
      help(argv[0]);
    else
```

```
      error(argv[0], argv[0], 1); /* Syntax error */
  else
    tmp = 1;
  dot = strlen(argv[tmp]);
  filename = malloc((dot + 5)*sizeof(char));
  if (filename == NULL)
    error(argv[0], argv[0], 4); /* Out of memory */
  strcpy(filename, argv[tmp]);

/* Load input data */

  strcat(filename, ".dat");
  fp = fopen(filename,"r");
  if (fp == NULL)
    error(argv[0], filename, 2);  /* File not found */
  fscanf(fp, "%lf %lf %lf", &ca, &cb, &cc);
  fscanf(fp, "%lf", &coff);
  fscanf(fp, "%d %d", &minup, &mindn);
  fscanf(fp, "%lf %lf", &cup, &cdn);
  fscanf(fp, "%lf %lf", &pmin, &pmax);
  fscanf(fp, "%lf %lf", &tmpd, &tmpd1);
  fscanf(fp, "%lf %lf %lf %lf", &eta, &bbar, &m, &sigma);
  b0 = log(tmpd) - m*tmpd1;
  if (prgopt == 1)
    b0 = bbar;
  fscanf(fp, "%d %d", &nd, &np);
  for (i = 0; i < np; i++)
    fscanf(fp, "%lf %lf", exld+i, stdld+i);
  fscanf(fp, "%lf", &disc);
  fclose(fp);

  filename[dot] = '\0';
  strcat(filename, ".rdt");
  fp = fopen(filename,"r");
  if (fp == NULL)
    error(argv[0], filename, 2);  /* File not found */
  fscanf(fp, "%lf %lf", &prr, &prf);
  fscanf(fp, "%lf %lf", &Kr, &sigr);
  fclose(fp);

  printf ("Enter hour of first decision (0-%d):",np);
  scanf ("%d", &hour);

/* Initialize dynamic programming */

    hlen = nd*np;
    tmpd = stdld[0];
    for (i = 1; i < np; i++)
      if (tmpd < stdld[i])
        tmpd = stdld[i];
    stdev = sqrt(sigma*sigma + m*m*tmpd*tmpd);
    nstdsc = ceil(3.5*sqrt(2)*stdev/disc);
    nsttot = 2*nstdsc*(hlen+1) + 1;
    iref = nstdsc*(hlen + 1);
    nst = minup + mindn;
    if (2*nstdsc+1 > PSTATES)
      error(argv[0], "1", 6); /* Not enough states for discretization level */
```

```
    if (nst*nsttot > STATES)
      error(argv[0], "1", 6); /* Not enough states for discretization level */
    Jc = Jk[0]; J1 = Jk[1];
    uc = uk[0]; u1 = uk[1];
    for (i = 0; i < nst*nsttot; i++)
      J1[i] = 0;
    filename[dot] = '\0';
    strcat(filename, ".dym");
    fp = fopen(filename,"w");
    if (fp == NULL)
      error(argv[0], argv[2], 3);    /* Unable to write to file */
    printf("Performing dynamic programming . . .\n");
    sq2 = sqrt(2);
    pmcmin = 2*ca*pmin + cb;
    pmcmax = 2*ca*pmax + cb;
    stlim = nstdsc*hlen;

/* Dynamic Programming */

    for (i = hlen; i >= 0; i--) {
      for (j = -stlim; j <= stlim; j++) {
        index = (i + hour) % np;
        ex = exp(-eta)*(b0 + j*disc - bbar) + bbar + m*exld[index];
        stdev = sqrt(sigma*sigma + m*m*stdld[index]*stdld[index]);
        expr = exp(ex + 0.5*stdev*stdev);
        zl = (log(pmcmin) - ex)/stdev;
        zu = (log(pmcmax) - ex)/stdev;
        extr = pmcmin*(0.5 + 0.5*erf(zl/sq2))
          + pmcmax*(0.5 - 0.5*erf(zu/sq2));
        extr = extr + 0.5*expr*(erf((zu - stdev)/sq2)
                                 - erf((zl - stdev)/sq2));
        vtr = 0.5*exp(2*ex+2*stdev*stdev)*(erf((zu-2*stdev)/sq2)
                                            - erf((zl-2*stdev)/sq2));
        vtr = vtr + pmcmin*pmcmin*(0.5 + 0.5*erf(zl/sq2));
        vtr = vtr + pmcmax*pmcmax*(0.5 - 0.5*erf(zu/sq2)) -extr*extr;
        exprod = 0.5*exp(2*ex+2*stdev*stdev)*(erf((zu-2*stdev)/sq2)
                                               - erf((zl-2*stdev)/sq2));
        exprod = exprod
          + pmcmin*exp(ex+0.5*stdev*stdev)*(0.5 + 0.5*erf((zl-stdev)/sq2));
        exprod = exprod
          + pmcmax*exp(ex+0.5*stdev*stdev)*(0.5 - 0.5*erf((zu-stdev)/sq2));
         exexper = exp(0.5*sigr*sigr);
        ecost[j+iref] = (1 - prf)*(1 - prr)*((extr*extr + vtr - cb*cb
                        + 2*cb*expr - 2*exprod)/(4*ca) + cc)
                        + prf*(exprod*(1 - exp(Kr)*exexper) - cb*expr)/(2*ca);
        ex = ex + Kr;
        stdev = sqrt(stdev*stdev + sigr*sigr);
        expr = exp(ex + 0.5*stdev*stdev);
        extr = pmcmin*(0.5 + 0.5*erf(zl/sq2))
          + pmcmax*(0.5 - 0.5*erf(zu/sq2));
        extr = extr + 0.5*expr*(erf((zu - stdev)/sq2)
                                 - erf((zl - stdev)/sq2));
        vtr = 0.5*exp(2*ex+2*stdev*stdev)*(erf((zu-2*stdev)/sq2)
                                            - erf((zl-2*stdev)/sq2));
        vtr = vtr + pmcmin*pmcmin*(0.5 + 0.5*erf(zl/sq2));
        vtr = vtr + pmcmax*pmcmax*(0.5 - 0.5*erf(zu/sq2)) -extr*extr;
        exprod = 0.5*exp(2*ex+2*stdev*stdev)*(erf((zu-2*stdev)/sq2)
```

```
                                            - erf((zl-2*stdev)/sq2));
    exprod = exprod
      + pmcmin*exp(ex+0.5*stdev*stdev)*(0.5 + 0.5*erf((zl-stdev)/sq2));
    exprod = exprod
      + pmcmax*exp(ex+0.5*stdev*stdev)*(0.5 - 0.5*erf((zu-stdev)/sq2));
    ecost[j+iref] = ecost[j+iref] + (1 - prf)*prr*((extr*extr + vtr
                      - cb*cb + 2*cb*expr - 2*exprod)/(4*ca) + cc)
                      + prf*(cb*expr)/(2*ca);
}
stdev = sqrt(sigma*sigma + m*m*stdld[index]*stdld[index]);
for (j = 0; j < nst; j++) {
  next1 = nextst(j,1,minup,mindn);
  next0 = nextst(j,0,minup,mindn);
  for (k = -stlim; k <= stlim; k++) {
    tmpd = stdev*sq2;
    ex = exp(-eta)*(b0 + k*disc - bbar) + bbar;
    tmpd1 = erf((b0 + (k - 0.5)*disc - ex)/tmpd);
    tmpd2 = erf((b0 + (k + 0.5)*disc - ex)/tmpd);
    probmat[nstdsc] = (tmpd2 - tmpd1)/2;
    for (k1 = 1; k1 <= nstdsc; k1++) {
      tmpd3 = erf((b0 + (k-k1-0.5)*disc - ex)/tmpd);
      tmpd4 = erf((b0 + (k+k1+0.5)*disc - ex)/tmpd);
      probmat[nstdsc+k1] = (tmpd4 - tmpd2)/2;
      probmat[nstdsc-k1] = (tmpd1 - tmpd3)/2;
      tmpd1 = tmpd3;
      tmpd2 = tmpd4;
    }
    if (j < minup) {
      lu = 1;
      lcost = ecost[k+iref];
      for (k1 = -nstdsc; k1 <= nstdsc; k1++) {
        index = nsttot*next1 + k1 + k + iref;
        lcost += probmat[k1+nstdsc]*J1[index];
      }
      if (j == minup - 1) {
        tmpd = coff + cdn;
        for (k1 = -nstdsc; k1 <= nstdsc; k1++) {
          index = nsttot*next0 + k1 + k + iref;
          tmpd += probmat[k1+nstdsc]*J1[index];
        }
        if (tmpd < lcost) {
          lu = 0;
          lcost = tmpd;
        }
      }
    }
    else {
      lu = 0;
      lcost = coff;
      for (k1 = -nstdsc; k1 <= nstdsc; k1++) {
        index = nsttot*next0 + k1 + k + iref;
        lcost += probmat[k1+nstdsc]*J1[index];
      }
      if (j == nst - 1) {
        tmpd = ecost[k+iref] + cup;
        for (k1 = -nstdsc; k1 <= nstdsc; k1++) {
          index = nsttot*next1 + k1 + k + iref;
```

```
                    tmpd += probmat[k1+nstdsc]*J1[index];
                  }
                  if (tmpd < lcost) {
                    lu = 1;
                    lcost = tmpd;
                  }
                }
              }
              index = nsttot*j + k + iref;
              Jc[index] = lcost;
              uc[index] = lu;
          }
        }
        stlim -= nstdsc;
        printf("%3d ... ", i);
        if (i == 1) {
          fprintf(fp,"%3d: ",i);
          for (j = 0; j < nst; j++) {
            for (k = -nstdsc; k <= nstdsc; k++) {
              index = j*nsttot + k + iref;
              fprintf(fp, "%7.2f ", Jc[index]);
              if ((k+nstdsc) % 6 == 5)
                fprintf(fp, "\n        ");
            }
            fprintf(fp,"\n        ");
          }
          fprintf(fp, "\n        ");
          for (j = 0; j < nst; j++) {
            for (k = -nstdsc; k <= nstdsc; k++) {
              index = j*nsttot + k + iref;
              fprintf(fp, "%10d ", uc[index]);
              if ((k+nstdsc) % 6 == 5)
                fprintf(fp, "\n        ");
            }
            fprintf(fp,"\n        ");
          }
          printf("\n(%d %d) ... ",stlim, abs(-nstdsc));
        }
        Jtmp = Jc; Jc = J1; J1 = Jtmp;
        utmp = uc; uc = uc; u1 = utmp;
      }
      fprintf(fp,"%3d: ",0);
      for (j = 0; j < nst; j++) {
        fprintf(fp, "%7.2f ", J1[j*nsttot + iref]);
        if (j % 6 == 5)
          fprintf(fp, "\n        ");
      }
      fprintf(fp,"\n        ");
      for (j = 0; j < nst; j++) {
        fprintf(fp, "%10d ", u1[j*nsttot + iref]);
        if (j % 6 == 5)
          fprintf(fp, "\n        ");
      }
      fprintf(fp,"\n        ");
      printf("\n");
      fclose(fp);
}
```

D.7   Source Code for `uci2con.c`

```
/*  Name:  Eric Allen
    Date:  3-24-98
    Description: Program to implement the dynamic programming algorithm for
    unit commitment for an individual power producer, using price process
    model with load forecasts and exponential algorithm.  Log of price is
    modeled as a mean-reverting intercept.  Simple congestion model is
    included.
*/

#include<stdio.h>
#include<math.h>
#include<stdlib.h>
#include<string.h>

#include"ucerr.h"          '
/* #include"uc1.h" */

#define EPSILON 0.0000001
#define PI 3.14159265358979

#define MAXP 96
#define TSTEPS 10
#define STATES 10000
#define PSTATES 500
#define ALT 3

double erf (double z) {

  double term, temp, fact, sum;
  int i;

  if (z <= -3.5)
    return (-1);
  if (z >= 3.5)
    return (1);

  i = 1;
  fact = 1;
  sum = z;

  do {
    fact = fact*i;
    temp = (2.0*i + 1);
    term = pow(z, temp)/ temp / fact;
    if (2 * (i/2) != i++)
      sum -= term;
    else
      sum += term;
  } while ((fabs(term) > EPSILON) && (i < 100));
  return (2*sum/sqrt(PI));
}

/* Calculates the next state from the current state and control option.
   There are minup + mindn possible states.  States 0, 1, ..., minup-1
   correspond to the generator having been up 1, 2, ..., minup periods, while
```

```
    states minup, minup+1, ..., minup+mindn-1 correspond to the generator
    having been down 1, 2, ..., mindn periods. */

int nextst(int currst, int opt, int mup, int mdn) {

  int n;

  if (currst < mup)
    if (opt == 0)
      n = mup;
    else if (currst < mup - 1)
      n = currst + 1;
    else
      n = currst;
  else
    if (opt)
      n = 0;
    else if (currst < mup + mdn - 1)
      n = currst + 1;
    else
      n = currst;
  return(n);
}

main(int argc, char *argv[])
{
  double ca, cb, cc;        /* cost curve coefficients */
  double coff;              /* expected cost per stage while off */
  int minup, mindn;         /* Minimum on/off times */
  double cup, cdn;          /* Startup/shutdown costs */
  double pmin, pmax;        /* Minimum/maximum generation levels */

  double exld[MAXP], stdld[MAXP]; /* Mean/variance of load forecast */

  double ex, stdev;         /* Mean/variance of price */
  double expr;              /* Expected price */
  double extr, vtr, sq2;    /* Mean/variance of truncated variable; sqrt(2) */
  double exprod;            /* Expected product of price and marginal cost */
  double pmcmin, pmcmax;    /* Marginal cost limits */
  double zl, zu;            /* Z- and Z+ (normalized limits) */

  double clim[MAXP], plim[MAXP];   /* Congestion limits and probabilities */
  double uplim;                    /* Upper generation limit (with cong.) */
  int numlim;                      /* Number of possible values of P_Clim */

  double Jk[2][STATES];            /* cost-to-go function */
  int uk[2][STATES];               /* optimal control */
  double probmat[PSTATES];         /* Transition probabilities */

  double *Jc, *J1, *Jtmp;          /* current and next state cost-to-go */
  int *uc, *u1, *utmp;             /* current and next state control */

  int index;                /* index array to find state no. */
  int state;                /* state counter */
  int hour;                 /* First hour in time horizon */

  double ecost[STATES];            /* Expected cost for one stage */
```

```
int lu;                          /* Minimizing control */
double lcost;                    /* Cost of minimizing control */

double b0;                       /* Starting intercept */
double eta, bbar, m, sigma;      /* Price process parameters */

double disc;                     /* Discretization level */
int nstdsc;                      /* No. of discretized states */
int nsttot;                      /* Total number of price states */
int iref;                        /* Reference into price state array */
int stlim;                       /* Price state limit for given stage */

int i, j, k, k1, nd, np, nst, hlen, nalt, nopt, baseopt, curropt, tmp, bit;
int next0, next1;                /* Next state for 0/1 choice of control */
double tmpd, tmpd1, tmpd2, tmpd3, tmpd4;
char check;

int prgopt = 0;                  /* Program options selected */
int dot;                         /* Length of filename (without extender) */
char *filename;
FILE *fp;

/* nd:      number of days in horizon
   np:      number of periods per day
   nst:     number of states
   nalt:    number of control alternatives
   nopt:    number of control options available for a given state
   baseopt: control alternative with no generator status changed
   curropt: current control alternative being examined
   hlen:    length of horizon
   bit:     bit mask to calculate control alternative number
   stlim:   limit of number of price states at current stage
*/

/* Process command line arguments */

  if ((argc < 2) || (argc > 3))
    error(argv[0], argv[0], 1);   /* Syntax error */
  if (argc == 3)
    if (strlen(argv[1]) != 2)
      error(argv[0], argv[0], 1); /* Syntax error */
    else if (argv[1][0] != '-')
      error(argv[0], argv[0], 1); /* Syntax error */
    else {
      switch(argv[1][1]) {
      case 'b': prgopt = 1; break;
      case 'h': help(argv[0]);
      case 'o': prgopt = 2; break;
      default: error(argv[0], argv[0], 1); /* Syntax error */
      }
      tmp = 2;
    }
  else if (argv[1][0] == '-')
    if (strlen(argv[1]) != 2)
      error(argv[0], argv[0], 1); /* Syntax error */
    else if (argv [1][1] == 'h')
      help(argv[0]);
```

```
      else
         error(argv[0], argv[0], 1); /* Syntax error */
    else
      tmp = 1;
   dot = strlen(argv[tmp]);
   filename = malloc((dot + 5)*sizeof(char));
   if (filename == NULL)
      error(argv[0], argv[0], 4); /* Out of memory */
   strcpy(filename, argv[tmp]);

/* Load input data */

   strcat(filename, ".dat");
   fp = fopen(filename,"r");
   if (fp == NULL)
      error(argv[0], filename, 2);   /* File not found */
   fscanf(fp, "%lf %lf %lf", &ca, &cb, &cc);
   fscanf(fp, "%lf", &coff);
   fscanf(fp, "%d %d", &minup, &mindn);
   fscanf(fp, "%lf %lf", &cup, &cdn);
   fscanf(fp, "%lf %lf", &pmin, &pmax);
   fscanf(fp, "%lf %lf", &tmpd, &tmpd1);
   fscanf(fp, "%lf %lf %lf %lf", &eta, &bbar, &m, &sigma);
   b0 = log(tmpd) - m*tmpd1;
   if (prgopt == 1)
      b0 = bbar;
   fscanf(fp, "%d %d", &nd, &np);
   for (i = 0; i < np; i++)
      fscanf(fp, "%lf %lf", exld+i, stdld+i);
   fscanf(fp, "%lf", &disc);
   fclose(fp);

   filename[dot] = '\0';
   strcat(filename, ".cdt");
   fp = fopen(filename,"r");
   if (fp == NULL)
      error(argv[0], filename, 2);   /* File not found */
   fscanf(fp, "%d", &numlim);
   for (i = 0; i < numlim; i++)
      fscanf(fp, "%lf %lf", clim+i, plim+i);
   fclose(fp);

   printf ("Enter hour of first decision (0-%d):",np);
   scanf ("%d", &hour);

/* Initialize dynamic programming */

      hlen = nd*np;
      tmpd = stdld[0];
      for (i = 1; i < np; i++)
         if (tmpd < stdld[i])
            tmpd = stdld[i];
      stdev = sqrt(sigma*sigma + m*m*tmpd*tmpd);
      nstdsc = ceil(3.5*sqrt(2)*stdev/disc);
      nsttot = 2*nstdsc*(hlen+1) + 1;
      iref = nstdsc*(hlen + 1);
      nst = minup + mindn;
```

```
      if (2*nstdsc+1 > PSTATES)
        error(argv[0], "1", 6); /* Not enough states for discretization level */
      if (nst*nsttot > STATES)
        error(argv[0], "1", 6); /* Not enough states for discretization level */
      Jc = Jk[0]; J1 = Jk[1];
      uc = uk[0]; u1 = uk[1];
      for (i = 0; i < nst*nsttot; i++)
        J1[i] = 0;
      filename[dot] = '\0';
      strcat(filename, ".dym");
      fp = fopen(filename,"w");
      if (fp == NULL)
        error(argv[0], argv[2], 3);     /* Unable to write to file */
      printf("Performing dynamic programming . . .\n");
      sq2 = sqrt(2);
      pmcmin = 2*ca*pmin + cb;
      pmcmax = 2*ca*pmax + cb;
      stlim = nstdsc*hlen;

/* Dynamic Programming */

      for (i = hlen; i >= 0; i--) {
        for (j = -stlim; j <= stlim; j++) {
          index = (i + hour) % np;
          ecost[j+iref] = 0;
          for (k = 0; k < numlim; k++) {
            ex = exp(-eta)*(b0 + j*disc - bbar) + bbar + m*exld[index];
            stdev = sqrt(sigma*sigma + m*m*stdld[index]*stdld[index]);
            expr = exp(ex + 0.5*stdev*stdev);
            zl = (log(pmcmin) - ex)/stdev;
            uplim = pmcmax;
            if (clim[k] < pmax)
              uplim = 2*ca*clim[k] + cb;
            zu = (log(uplim) - ex)/stdev;
            extr = pmcmin*(0.5 + 0.5*erf(zl/sq2))
              + uplim*(0.5 - 0.5*erf(zu/sq2));
            extr = extr + 0.5*expr*(erf((zu - stdev)/sq2)
                                      - erf((zl - stdev)/sq2));
            vtr = 0.5*exp(2*ex+2*stdev*stdev)*(erf((zu-2*stdev)/sq2)
                                      - erf((zl-2*stdev)/sq2));
            vtr = vtr + pmcmin*pmcmin*(0.5 + 0.5*erf(zl/sq2));
            vtr = vtr + uplim*uplim*(0.5 - 0.5*erf(zu/sq2)) -extr*extr;
            exprod = 0.5*exp(2*ex+2*stdev*stdev)*(erf((zu-2*stdev)/sq2)
                                      - erf((zl-2*stdev)/sq2));
            exprod = exprod
              + pmcmin*exp(ex+0.5*stdev*stdev)*(0.5 + 0.5*erf((zl-stdev)/sq2));
            exprod = exprod
              + uplim*exp(ex+0.5*stdev*stdev)*(0.5 - 0.5*erf((zu-stdev)/sq2));
            ecost[j+iref] = ecost[j+iref]
              + plim[k]*((extr*extr + vtr - cb*cb + 2*cb*expr
                        - 2*exprod)/(4*ca) + cc);
          }
        }
        for (j = 0; j < nst; j++) {
          next1 = nextst(j,1,minup,mindn);
          next0 = nextst(j,0,minup,mindn);
          for (k = -stlim; k <= stlim; k++) {
```

```
        tmpd = stdev*sq2;
        ex = exp(-eta)*(b0 + k*disc - bbar) + bbar;
        tmpd1 = erf((b0 + (k - 0.5)*disc - ex)/tmpd);
        tmpd2 = erf((b0 + (k + 0.5)*disc - ex)/tmpd);
        probmat[nstdsc] = (tmpd2 - tmpd1)/2;
        for (k1 = 1; k1 <= nstdsc; k1++) {
          tmpd3 = erf((b0 + (k-k1-0.5)*disc - ex)/tmpd);
          tmpd4 = erf((b0 + (k+k1+0.5)*disc - ex)/tmpd);
          probmat[nstdsc+k1] = (tmpd4 - tmpd2)/2;
          probmat[nstdsc-k1] = (tmpd1 - tmpd3)/2;
          tmpd1 = tmpd3;
          tmpd2 = tmpd4;
        }
        if (j < minup) {
          lu = 1;
          lcost = ecost[k+iref];
          for (k1 = -nstdsc; k1 <= nstdsc; k1++) {
            index = nsttot*next1 + k1 + k + iref;
            lcost += probmat[k1+nstdsc]*J1[index];
          }
          if (j == minup - 1) {
            tmpd = coff + cdn;
            for (k1 = -nstdsc; k1 <= nstdsc; k1++) {
              index = nsttot*next0 + k1 + k + iref;
              tmpd += probmat[k1+nstdsc]*J1[index];
            }
            if (tmpd < lcost) {
              lu = 0;
              lcost = tmpd;
            }
          }
        }
        else {
          lu = 0;
          lcost = coff;
          for (k1 = -nstdsc; k1 <= nstdsc; k1++) {
            index = nsttot*next0 + k1 + k + iref;
            lcost += probmat[k1+nstdsc]*J1[index];
          }
          if (j == nst - 1) {
            tmpd = ecost[k+iref] + cup;
            for (k1 = -nstdsc; k1 <= nstdsc; k1++) {
              index = nsttot*next1 + k1 + k + iref;
              tmpd += probmat[k1+nstdsc]*J1[index];
            }
            if (tmpd < lcost) {
              lu = 1;
              lcost = tmpd;
            }
          }
        }
        index = nsttot*j + k + iref;
        Jc[index] = lcost;
        uc[index] = lu;
      }
    }
    stlim -= nstdsc;
```

```
    printf("%3d ... ", i);
    if (i == 1) {
      fprintf(fp,"%3d: ",i);
      for (j = 0; j < nst; j++) {
        for (k = -nstdsc; k <= nstdsc; k++) {
          index = j*nsttot + k + iref;
          fprintf(fp, "%7.2f ", Jc[index]);
          if ((k+nstdsc) % 6 == 5)
            fprintf(fp, "\n      ");
        }
        fprintf(fp,"\n      ");
      }
      fprintf(fp, "\n      ");
      for (j = 0; j < nst; j++) {
        for (k = -nstdsc; k <= nstdsc; k++) {
          index = j*nsttot + k + iref;
          fprintf(fp, "%10d ", uc[index]);
          if ((k+nstdsc) % 6 == 5)
            fprintf(fp, "\n      ");
        }
        fprintf(fp,"\n      ");
      }
      printf("\n(%d %d) ... ",stlim, abs(-nstdsc));
    }
    Jtmp = Jc; Jc = J1; J1 = Jtmp;
    utmp = uc; uc = uc; u1 = utmp;
  }
  fprintf(fp,"%3d: ",0);
  for (j = 0; j < nst; j++) {
    fprintf(fp, "%7.2f ", J1[j*nsttot + iref]);
    if (j % 6 == 5)
      fprintf(fp, "\n      ");
  }
  fprintf(fp,"\n      ");
  for (j = 0; j < nst; j++) {
    fprintf(fp, "%10d ", u1[j*nsttot + iref]);
    if (j % 6 == 5)
      fprintf(fp, "\n      ");
  }
  fprintf(fp,"\n      ");
  printf("\n");
  fclose(fp);
}
```

D.8   Source Code for `ucerr.h`

```
/* This function is the error processing routine.  It receives pointers to
   the strings containing the program name and file name and an error code.
   The error codes are:

      1:  Syntax error
      2:  File not found
      3:  Unable to write to file
      4:  Out of memory
      5:  Input .dat file format error
      6:  Discretization too fine

*/

void error(char *prgname, char *fname, int code) {

  switch(code) {
  case 1: printf ("Usage: %s [-option] file\n", prgname);
    printf ("Type '%s -h | more' for complete help\n\n", prgname);
    break;
  case 2: printf ("%s: Unable to find %s\n", prgname, fname);
    break;
  case 3: printf ("%s: Unable to open %s\n", prgname, fname);
    break;
  case 4: printf ("%s: Out of memory\n", prgname);
    break;
  case 6: printf ("%s: Discretization too small\n", prgname);
    break;
  }
  exit(EXIT_FAILURE);
}

/* This subroutine prints the help message, actuated by option '-h' */

void help(char *prgname) {

  printf ("Usage: %s [-option] file\n\n", prgname);
  printf ("%s is the unit commitment scheduling program.  The input data\n",
          prgname);
  printf ("is contained in 'file.dat' and the output is contained in \n");
  printf ("'file.dym'.  The available options, only one of which may be\n");
  printf ("selected, are:\n\n");
  printf ("   -b   Set initial price intercept to its mean value\n\n");
  printf ("   -h   Print help message\n\n");
  exit(EXIT_SUCCESS);
}
```

D.9   Source Code for `ucerroo.h`

```
/* This function is the error processing routine.  It receives pointers to
   the strings containing the program name and file name and an error code.
   The error codes are:

      1:  Syntax error
      2:  File not found
      3:  Unable to write to file
      4:  Out of memory
      5:  Input .dat file format error
      6:  Discretization too fine

*/

void error(char *prgname, char *fname, int code) {

  switch(code) {
  case 1: printf ("Usage: %s [-option] file\n", prgname);
    printf ("Type '%s -h | more' for complete help\n\n", prgname);
    break;
  case 2: printf ("%s: Unable to find %s\n", prgname, fname);
    break;
  case 3: printf ("%s: Unable to open %s\n", prgname, fname);
    break;
  case 4: printf ("%s: Out of memory\n", prgname);
    break;
  case 6: printf ("%s: Out of array space\n", prgname);
    break;
  }
  exit(EXIT_FAILURE);
}

/* This subroutine prints the help message, actuated by option '-h' */

void help(char *prgname) {

  printf ("Usage: %s [-option] file\n\n", prgname);
  printf ("%s is the unit commitment scheduling program.  The input data\n",
          prgname);
  printf ("is contained in 'file.dat' and 'file.par' and the output is\n");
  printf ("contained in 'file.oo'.  The available options, only one of\n");
  printf ("which may be selected, are:\n\n");
  printf ("   -b   Set initial price intercept to its mean value\n\n");
  printf ("   -h   Print help message\n\n");
  exit(EXIT_SUCCESS);
}
```

APPENDIX E
STOCHASTIC UNIT COMMITMENT IN A REGULATED INDUSTRY

The theory of dynamic programming can also be applied to the stochastic unit commitment problem in a regulated environment. This appendix illustrates some methods for computing the expected cost per stage for this problem. As compared to previous work noted in Chapter 2, this appendix considers the total load as a random quantity and also takes into account possible generator failures. Power levels for each generator are determined by optimal power flow.

E.1 Unit Commitment with a Finite Horizon

For the unit commitment problem, the control decision consists of choosing which generators will be run in order to meet the load and reserve requirements during that stage. The system state vector simply indicates how long each generator has been continuously on or off; this information poses constraints on the control decision in the form of minimum on and off times and possibly maximum run times for each generator. Scheduled maintenance outages for certain units may also limit the control choices available at some stages. We will use the same convention as [3]; if $x_k(i) > 0$, then generator i has been up for $x_k(i)$ stages; otherwise, generator i has been down for $-x_k(i)$ stages. The state transition equation is [3]:

$$x_{k+1}(i) = \begin{cases} \max(1, x_k(i) + 1) & : u_k(i) = 1 \\ \min(-1, x_k(i) - 1) & : u_k(i) = 0 \end{cases} \tag{E.1}$$

Of course, the real and reactive power demand at each bus at any time is not known in advance and must be treated as a random variable. The availability of any generator and transmission line is also unknown, as there is always the possibility that a component might fail. The optimal unit commitment decision is to minimize the total expected cost over all stages in the presence of these uncertain quantities [1]. Note that previous DP formulations of unit commitment [3] have treated all quantities as known; here we allow for the presence of unknown variables, changing the DP problem from a deterministic problem to a stochastic one.

E.2   Solution of Optimal Power Flow

Recall that the unit commitment cost requires the calculation of the optimal power flow injections for a random demand. In this section, we will perform the analytic evaluation of the optimal power flow minimization in equation (2.15).

Throughout this book, we will assume that the cost of generation is a quadratic function of the power output:

$$c_{Gi}(P_{Gi}) = a_i P_{Gi}^2 + b_i P_{Gi} + c_i \tag{E.2}$$

Furthermore, we denote by $G_{\mathbf{u}}$ the set of generators which are turned on for the control choice $\mathbf{u} = [u(1)u(2)\cdots u(N_G)]^T$.

In calculating the optimal power flow, we first consider a simplified system with no generation limits, losses, or congestion. We will then examine how these constraints modify the solution to the simplified problem.

E.2.1   Uncongested, Lossless Optimal Power Flow without Generation Limits

For this case, with quadratic cost curves, an exact analytic solution can be obtained. We first note that the only constraint equation for the system is that total power generated equals total demand:

$$P_L = \sum_{i=1}^{N_L} P_{Li} = \sum_{i \in G_{\mathbf{u}}} P_{Gi} \tag{E.3}$$

while the objective is to minimize the total generation cost $\sum_{i \in G_{\mathbf{u}}} c_{Gi}(P_{Gi})$. Following [41], the Lagrangian (L) is:

$$L = \sum_{i \in G_{\mathbf{u}}} c_{Gi}(P_{Gi}) + \lambda(P_L - \sum_{i \in G_{\mathbf{u}}} P_{Gi}) \tag{E.4}$$

Substituting for the cost function, we have:

$$L = \sum_{i \in G_u} (a_i P_{Gi}^2 + b_i P_{Gi} + c_i) + \lambda(P_L - \sum_{i \in G_u} P_{Gi}) \tag{E.5}$$

Taking partial derivatives and setting them equal to zero, we find:

$$\frac{\partial L}{\partial \lambda} = P_L - \sum_{i \in G_u} P_{Gi} = 0 \tag{E.6}$$

$$\frac{\partial L}{\partial P_{Gi}} = 2a_i P_{Gi} + b_i - \lambda = 0 \tag{E.7}$$

The optimum is the solution point of the preceding system of $m+1$ equations, where m is the number of active generators for control $\mathbf{u}$. The unknowns are $P_{G(i)}$ for λ and all i from 1 to m, where $P_{G(i)}$ is the output power for the ith generator in $G_\mathbf{u}$. In matrix form, the optimal power flow problem is [20]:

$$\begin{bmatrix} 2a_1 & 0 & \cdots & 0 & -1 \\ 0 & 2a_2 & \cdots & 0 & -1 \\ \vdots & \vdots & \ddots & \vdots & \vdots \\ 0 & 0 & \cdots & 2a_m & -1 \\ 1 & 1 & \cdots & 1 & 0 \end{bmatrix} \begin{bmatrix} P_{G(1)} \\ P_{G(2)} \\ \vdots \\ P_{G(m)} \\ \lambda \end{bmatrix} = \begin{bmatrix} -b_1 \\ -b_2 \\ \vdots \\ -b_m \\ P_L \end{bmatrix} \tag{E.8}$$

Because of the special form of the matrix, it can be inverted in $O(n^2)$ operations. The resulting solution for P_{Gi} is a linear function of the demand:

$$P_{Gi} = e_{1i} P_L + e_{2i} \tag{E.9}$$

This result is very useful since it means that for a given configuration of active generators, a complete optimal power flow solution can be obtained by a single calculation of $O(n^2)$ operations. The e_{1i} and e_{2i} coefficients from this calculation can then be stored and used to determine the optimal generation level for any value of load demand. Of course, the generation cannot be negative; a negative P_{Gi} means that the marginal cost for that generator at zero is too high and therefore the generator should not be run for the given load value.

E.3   Expected Value of the Generation Cost

The dynamic programming procedure requires the calculation of the expected value of the generation cost. There are two basic random quantities to be considered: the power at the loads and the availability of generators and transmission lines. To calculate the expected cost, we will deal with these quantities separately.

E.3.1   Expected Cost over Load Power

For now, we will take as a given that the active generators will not fail. Assuming that the expected value and variance of the load are known (or can be estimated), it is a straightforward procedure to derive the expected value of the generation cost. Substituting equation (E.9) into the cost functions, the total generation cost becomes:

$$\sum_{i \in G_u} c_{Gi}(P_{Gi}) = \sum_{i \in G_u} \left[a_i e_{1i}^2 P_L^2 + (2 a_i e_{1i} e_{2i} + b_i e_{1i}) P_L + a_i e_{2i}^2 + b_i e_{2i} + c_i \right]$$

$$(E.10)$$

Taking the expected value with respect to the random variable P_L:

$$\underset{P_L}{E} \left\{ \sum_{i \in G_u} c_{Gi}(P_{Gi}) \right\} = \sum_{i \in G_u} \left[a_i e_{1i}^2 \underset{P_L}{E}\{P_L^2\} + (2 a_i e_{1i} e_{2i} + b_i e_{1i}) \underset{P_L}{E}\{P_L\} \right.$$
$$\left. + a_i e_{2i}^2 + b_i e_{2i} + c_i \right]$$

$$(E.11)$$

To compute $E\{P_L^2\}$, note that if the standard deviation of P_L is σ_L, then the variance is:

$$\sigma_L^2 = \underset{P_L}{E} \{(P_L - \overline{P}_L)^2\}$$

$$(E.12)$$

where $\overline{P}_L = E\{P_L\}$. Expanding the squared term and applying the expectation operator to each individual term:

$$\sigma_L^2 = \underset{P_L}{E}\{P_L^2\} - 2\overline{P}_L \underset{P_L}{E}\{P_L\} + \overline{P}_L^2$$

$$(E.13)$$

since $\overline{P}_L$ is a constant and not a random variable. Finally, solving for $E\{P_L^2\}$:

$$\underset{P_L}{E}\{P_L^2\} = \overline{P}_L^2 + \sigma_L^2$$

$$(E.14)$$

We now substitute this result back into equation (E.11) to obtain:

$$\underset{P_L}{E} \left\{ \sum_{i \in G_u} c_{Gi}(P_{Gi}) \right\} = \sum_{i \in G_u} \left[a_i e_{1i}^2 \overline{P}_L^2 + (2 a_i e_{1i} e_{2i} + b_i e_{1i}) \overline{P}_L \right.$$
$$\left. + a_i e_{2i}^2 + b_i e_{2i} + c_i + a_i e_{1i}^2 \sigma_L^2 \right]$$

$$(E.15)$$

For simplicity, equation (E.15) may be written as:

$$\underset{P_L}{E} \left\{ \sum_{i \in G_u} c_{Gi}(P_{Gi}) \right\} = (C|\mathbf{u}) = e_{C2[\mathbf{u}]} \overline{P}_L^2 + e_{C1[\mathbf{u}]} \overline{P}_L + e_{C0[\mathbf{u}]}$$

$$(E.16)$$

with the definitions:

$$e_{C2[\mathbf{u}]} = \sum_{i \in G_{\mathbf{u}}} a_i e_{1i}^2 \qquad (E.17)$$

$$e_{C1[\mathbf{u}]} = \sum_{i \in G_{\mathbf{u}}} (2a_i e_{1i} e_{2i} + b_i e_{1i}) \qquad (E.18)$$

$$e_{C0[\mathbf{u}]} = \sum_{i \in G_{\mathbf{u}}} (a_i e_{2i}^2 + b_i e_{2i} + c_i + a_i e_{1i}^2 \sigma_L^2) \qquad (E.19)$$

E.3.2 Expected Cost over Generation Failures

We will now modify the expected generation cost to account for the possibility of generator outages. We ignore line outages for now; for a lossless system without congestion, line outages will not affect the ability to serve any load unless all of the lines connecting a bus fail simultaneously, which in general is highly improbable. We also assume that demand and generator failures are independent random variables.

In order to precisely calculate the expected cost accounting for generator availability, we need to perform an optimal power flow calculation for every subset of generators from the set of active generators. For n generators, there are 2^n subsets, so clearly a complete calculation is not practical. We will instead perform the optimal power flow calculation only for systems with at most l_f generators failed; l_f will be referred to as the failure level. We will approximate the cost of failures by modeling each failure as occurring at the beginning of the stage, immediately following the control decision. Using this model, the expected generation cost during a single stage by using control $\mathbf{u}$ is:

$$
\begin{aligned}
C_{\mathbf{u}} \;=\; & (C|\mathbf{u}) \prod_{i \in G_{\mathbf{u}}} (1 - p_{fi}) \\
& + \sum_{i=1}^{l_f} \sum_{J:i \in G_{\mathbf{u}}} \left((C|\mathbf{u}_{fJ}) + \sum_{k \in J} T_k \right) \prod_{k \in J} p_{fk} \prod_{k \in (G_{\mathbf{u}} \cap \bar{J})} (1 - p_{fk}) \\
& + p_{MF} \sum_{i=1}^{N_L} I_{Li} \qquad (E.20)
\end{aligned}
$$

The symbol:

$$\sum_{I:n \in S}$$

is for a combinatorial sum. The set I takes on all subsets of S that have n elements; the number of terms in the summation is $\binom{s}{n}$, where s is the number of elements in S. N_{Gu} is the number of generators that are on for control $\mathbf{u}$. p_{fi} is the probability that generator i will fail during one stage; p_{MF} is the probability that more than l_f generators will fail during a single stage; this probability is computed as:

$$p_{MF} = 1 - \sum_{i=l_f+1}^{N_{Gu}} \sum_{J:i\in G_u} \prod_{k\in J} p_{fk} \prod_{k\in(G_u\cap\bar{J})} (1 - p_{fk}) \qquad (E.21)$$

The complementary set $\bar{J}$ is the set of all generators which are not in J. The control $\mathbf{u}_{fJ}$ is equal to $\mathbf{u}$ except that the generators in the set J are to be turned off. Mathematically:

$$u_{fJ}(i) = \left\{ \begin{array}{ll} 0 & : i \in J \\ u(i) & : i \notin J \end{array} \right. \qquad (E.22)$$

Equation (E.20) looks horribly complicated, but it simply states that the expected one-stage cost (not including generator startup or shutdown costs) is calculated by multiplying the probability of a specific set of generators failing (the product terms) by the sum of the shutdown costs for the set of generators and the expected cost per stage given that that particular set of generators is down and the other generators are up (the quantity $(C|\mathbf{u}_{fJ})$). This quantity is summed over all sets of possible failures of at most l_f generators. Failures of more than l_f generators are approximated by simply assuming that no load is served, so that all loads must receive the pre-specified insurance payment. Equation (E.20) is based on the law of alternatives from probability theory [42].

If interruptible service contracts are present, then the calculation of $(C|\mathbf{u}_{fJ})$ is made by finding the mean and variance of the total load that is not interrupted when the generators in J fail. Recall that the rationing of a load is associated with a subset of all possible combinations of generator failures.

BIBLIOGRAPHY

[1] D. P. Bertsekas, *Dynamic Programming and Optimal Control: Volume I.* Belmont, MA: Athena Scientific, 1995.

[2] D. Bertsekas, G. Lauer, N. Sandell, Jr., and T. Posbergh, "Optimal short-term scheduling of large-scale power systems," *IEEE Transactions on Automatic Control,* Vol. AC-28, No. 1, January 1983, pp. 1-11.

[3] J. J. Shaw, "A direct method for security-constrained unit commitment," *IEEE Transactions on Power Systems,* Vol. 10, No. 3, August 1995, pp. 1329-1342.

[4] S. Erwin et al, "Using an optimization software to lower overall electric production costs for Southern Company," *Interfaces,* Vol. 21, No. 1, January-February 1991, pp. 27-41.

[5] Citizens Power, *The US Power Market: Restructuring and Risk Management,* London: Risk Publications, 1997.

[6] A. Dixit and R. Pindyck, *Investment under Uncertainty,* Princeton, NJ: Princeton University Press, 1993.

[7] Electric Power Research Institute, "Option Pricing for Project Evaluation: An Introduction," TR-104755, Research Project 1920-05, Final Report, January 1995.

[8] M. Kawai, "Spot and futures prices of nonstorable commodities under rational expectations," *Quarterly Journal of Economics,* Vol. 98, No. 2, May 1983, pp. 235-254.

[9] S. Jabbour, G. Garman, and B. Louks, "Impacts of dynamic costs and benefits on utility operating and planning decisions," IEEE Power Engineering Society Summer Meeting, Long Beach, CA, July 9-14, 1989, No. 89 SM 613-1 EC.

[10] W. W. Hogan, "Contract networks for electric power transmission," *Journal of Regulatory Economics,* Vol. 4, 1992, pp. 211-242.

[11] S. Oren, P. Spiller, P. Varaiya, and F. Wu, "Nodal prices and transmission rights: a critical appraisal," *The Electricity Journal*, April 1995, pp. 24-35.

[12] S. Oren, "Economic inefficiency of passive transmission rights in congested electricity systems with competitive generation," *The Energy Journal*, Vol. 18, No. 1, 1997, pp. 62-83.

[13] Y. C. Ho, "Soft optimization for hard problems," computerized lecture via private communication/distribution.

[14] T. W. Lau and Y. C. Ho, "Universal alignment probabilities and subset selection for ordinal optimization," *Journal of Optimization Theory and Applications*, Vol. 93, No. 3, June 1997, pp. 455-489.

[15] J. Gruhl, F. Schweppe, and M. Ruane, "Unit commitment scheduling of electric power systems," *Proceedings of the Systems Engineering for Power: Status and Prospects*, Henniker, NH, 1972, pp. 116-128.

[16] S. Takriti, J. Birge, and E. Long, "A stochastic model for the unit commitment problem," *IEEE Transactions on Power Systems*, Vol. 11, No. 3, August 1996. pp. 1497-1506.

[17] M. Pereira and N. Balu, "Composite generation/transmission reliability evaluation," *Proceedings of the IEEE*, Vol. 80, No. 4, April 1992, pp. 469-491.

[18] H. Laffaye, P. Clavel, and C. Trzpit, "GEODE: a new design of the operations planning system at EDF," *Proceedings of the PSCC*, 1990, pp. 646-653.

[19] G. Douard and L. Feltin, "Reliability assessment tools: EDF's experience and projects."

[20] K. Hara, M. Kimura, and N. Honda, "A method for planning economic unit commitment and maintenance of thermal power systems," *IEEE Transactions on Power Apparatus and Systems*, Vol. PAS-85, No. 5, May 1966, pp. 427-436.

[21] North American Electric Reliability Council, "NERC Policy on Generation Control and Performance," 1997.

[22] C.-W. Tan and P. Varaiya, "Interruptible electric power service contracts," *Journal of Economic Dynamics and Control*, Vol. 17, 1993, pp. 495-517.

[23] F. Wu and P. Varaiya, "Coordinated multilateral trades for electric power networks: theory and implementation," Report PWP-031, University of California Energy Institute, June 1995.

[24] R. Johnson, S. Oren, and A. Svoboda, "Volatility of unit commitment in competitive electricity markets," *Proceedings of the 30th Annual Hawaii International Conference on System Sciences; Vol. 5: Advanced Technology Track,* Wailea, HI, January 30, 1997, pp. 594-601.

[25] Discussion with Pravin Varaiya, November 1996.

[26] E. H. Allen and M. D. Ilić, "Stochastic unit commitment in a deregulated utility industry," *The Proceedings of the 29th Annual North American Power Symposium,* Laramie, WY, October 13-14, 1997, pp. 105-112.

[27] PJM Interconnection LLC, World Wide Web site, http://www.pjm.com, 1997.

[28] D. Dickey and W. Fuller, "Likelihood ratio statistics for autoregressive time series with a unit root," *Econometrica,* Vol. 49, No. 4, July 1981, pp. 1057-1072.

[29] S. Weisberg, *Applied Linear Regression,* New York: John Wiley & Sons, Inc., 1980.

[30] F. Wolak, "Market design and price behavior in restructured electricity markets: an international comparison," http://www-leland.stanford.edu/~wolak, 1997.

[31] L. Ljung and T. Söderström, *Theory and Practice of Recursive Identification,* (MIT Press Series in Signal Processing, Optimization, and Control, No. 4), Cambridge, MA: MIT Press, 1983.

[32] National Climatic Data Center, National Oceanic and Atmospheric Administration, U.S. Department of Commerce, World Wide Web site, http://www.ncdc.noaa.gov, 1997.

[33] R. Kaye, H. Outhred, and C. Bannister, "Forward contracts for the operation of an electricity industry under spot pricing," *IEEE Transactions on Power Systems,* Vol. 5, No. 1, February 1990, pp. 46-52.

[34] T. Gedra and P. Varaiya, "Markets and pricing for interruptible electric power," *IEEE Transactions on Power Systems,* Vol. 8, No. 1, February 1993, pp. 122-128.

[35] M. Ilić, L. Hyman, E. Allen, R. Cordero, and C.-N. Yu, "Interconnected system operations and expense planning in a changing industry: coordination vs. competition," *Topics in Regulatory Economics and Policy Series,* Kluwer Academic Publishers, 1997, pp. 307-332.

[36] C.N.-Yu, "Congestion management pricing and control for a deregulated electric power industry," Ph.D. thesis, Massachusetts Institute of Technology, expected August 1998.

[37] A. R. Gallant, *Nonlinear Statistical Models,* New York: John Wiley & Sons, Inc., 1987.

[38] D. Ratkowsky, *Nonlinear Regression Modeling: A Unified Practical Approach,* New York: Marcel Dekker, Inc., 1983.

[39] R. L. Burden and J. D. Faires, *Numerical Analysis,* 4th ed., Boston, MA: PWS-KENT Publishing Company, 1989.

[40] M. Dahleh, 6.241 class notes, Massachusetts Institute of Technology, Fall Semester 1994.

[41] A. El-Abiad, *Power Systems Analysis and Planning,* New York: McGraw-Hill, 1983, pp. 47-62.

[42] Gian-Carlo Rota in collaboration with Kenneth Baclawski, Sara Billey, and Graham Sterling, *Introduction to Probability Theory.* Birkhäuser, 1995.

[43] S. M. Ross, *Introduction to Probability Models,* 6th ed., San Diego, CA: Academic Press, 1997.

INDEX